STELLAR ASTROPHYSICS WITH THE WORLD'S LARGEST TELESCOPES

Related Titles from AIP Conference Proceedings

745 International Symposium on High Energy Gamma-Ray Astronomy
Edited by F. A. Aharonian, H. J. Völk, and D. Horns, February 2005, CD-ROM included, 0-7354-0229-9

727 Gamma-Ray Bursts: 30 Years of Discovery: Gamma-Ray Burst Symposium
Edited by E. E. Fenimore and M. Galassi, September 2004, 0-7354-0208-6

714 X-Ray Timing 2003: Rossi and Beyond
Edited by Philip Kaaret, Frederick K. Lamb, and Jean H. Swank, July 2004, 0-7354-0193-4

713 The Search for Other Worlds: Fourteenth Astrophysics Conference
Edited by Stephen S. Holt and Drake Deming, June 2004, 0-7354-0190-X

666 The Emergence of Cosmic Structure: Thirteenth Astrophysics Conference
Edited by Stephen S. Holt and Christopher S. Reynolds, May 2003, 0-7354-0128-4

662 Gamma-Ray Burst and Afterglow Astronomy 2001: A Workshop Celebrating the First Year of the HETE Mission
Edited by G. R. Ricker and R. K. Vanderspek, April 2003, 0-7354-0122-5

637 Classical Nova Explosions: International Conference on Classical Nova Explosions
Edited by Margarita Hernanz and Jordi José, November 2002, 0-7354-0092-X

616 Experimental Cosmology at Millimetre Wavelengths: 2K1BC Workshop
Edited by Marco DePetris and Massimo Gervasi, May 2002, 0-7354-0062-8

609 Astrophysical Polarized Backgrounds: Workshop on Astrophysical Polarized Backgrounds
Edited by Stefano Cecchini, Stefano Cortiglioni, Robert Sault, and Carla Sbarra, March 2002, 0-7354-0055-5

599 X-Ray Astronomy: Stellar Endpoints, AGN, and the Diffuse X-Ray Background
Edited by Nicholas E. White, December 2001, 0-7354-0043-1

598 Solar and Galactic Composition: A Joint SOHO/ACE Workshop
Edited by Robert F. Wimmer-Schweingruber, December 2001, CD-ROM included, 0-7354-0042-3

587 Gamma 2001: Gamma-Ray Astrophysics 2001
Edited by Steven Ritz, Neil Gehrels, and Chris R. Shrader, October 2001, 0-7354-0027-X
CD-ROM: 0-7354-0030-X

565 Young Supernova Remnants: Eleventh Astrophysics Conference
Edited by Stephen S. Holt and Una Hwang, May 2001, 0-7354-0001-6

STELLAR ASTROPHYSICS WITH THE WORLD'S LARGEST TELESCOPES

First International Workshop on Stellar Astrophysics with the World's Largest Telescopes

Torun, Poland 7 – 10 September 2004

EDITORS
Joanna Mikołajewska
Nicolaus Copernicus Astronomical Center
Warsaw, Poland
Arkadiusz Olech
Nicolaus Copernicus Astronomical Center
Warsaw, Poland

SPONSORING ORGANIZATIONS
Polish Ministry of Scientific Research and Information Technology
Nicolaus Copernicus Astronomical Center of Polish Academy of Sciences

AMERICAN INSTITUTE OF PHYSICS

Melville, New York, 2005
AIP CONFERENCE PROCEEDINGS ■ VOLUME 752

Editors:

Joanna Mikołajewska
Arkadiusz Olech

Nicolaus Copernicus Astronomical Center
Bartycka 18
00-716 Warsaw
Poland

E-mail: mikolaj@camk.edu.pl
olech@camk.edu.pl

Cover: Courtesy of Emilio Segrè Visual Archives, American Institute of Physics, College Park, MD

L.C. Catalog Card No. 2005921809

ISBN 0-7354-0237-X
ISSN 0094-243X
Printed in the United States of America

CONTENTS

END PRODUCTS OF SINGLE AND BINARY STELLAR EVOLUTION

LOCAL GROUP OF GALAXIES

CONCLUDING REMARKS

APPENDICES

PREFACE

A new generation of large, ground-based telescopes (e.g. the 10m Hobby-Eberly and Keck, or the 8m VLT and Gemini) have come into operation in the recent years. This has created new challenges for the astronomical community; in particular, to be able to develop modern and successful instrumentation for those telescopes, and to find the most important problems for them to solve. An important new addition to the family of very large telescopes is the Southern African Large Telescope (SALT), built together by Poland, Germany, UK, USA, South Africa, and New Zealand. It has been completed by the end of 2004, becoming the largest single telescope in the southern hemisphere. The access to one of the world largest telescopes is a great challenge for astronomers from both Poland and the other SALT countries, who will get unprecedented new opportunities for their research. This development also creates a strong need for their training both in modern astrophysics and in developing and using modern technological tools. So far, most of the research conducted by Polish astronomers (often in collaboration with both East- and West-bound astronomers) has been in the field of stellar astrophysics. This workshop, held in Poland and dedicated to key problems of stellar astrophysics that could and should be addressed by the users of these large telescopes, was considered by us as a very timely occasion to strengthen the links and collaborations between the researchers from different countries. Furthermore, a goal of this workshop, first ever on such a subject, was to direct our future research toward important issues in the modern ground-based observational astronomy.

The workshop was held 2004 September 7-10 in the Gromada Hotel in Toruń, a delightful medieval town on the UNESCO list of World Heritage Sites, also famous for being the birthplace of Nicolaus Copernicus (born in 1473). It was attended by about 50 astronomers from 15 different countries working in a number of fields related in various ways to the stellar astrophysics. The scientific program included 13 invited reviews and 16 contributed talks, which have covered the most timely and interesting problems of stellar astrophysics suitable for studies with the world largest telescopes. Among the most important achievements of the workshop was a comprehensive presentation of the state-of-the-art research in stellar and planetary system formation, extra-solar planets, final stages of single and binary stellar evolution, and stellar populations in the Local Group of Galaxies, including observational techniques and technologies applicable to those important fields. These and other topics addressed during the meeting will hopefully help to identify directions for future research, especially given that much attention was paid to identify the most important unresolved problems of stellar astrophysics. We also hope that the workshop has provided a friendly environment for discussions and will promote future research projects involving collaboration between different participants' institutions. Finally, we believe that these proceedings, containing the invited reviews and contributed talks presented during the workshop, will provide a useful introduction to these studies.

We would like to take this opportunity to thank all the people involved in the organization of the workshop, in particular the members of the Scientific Organizing Committee, for their assistance in preparing the scientific program. We also thank the participants for their active participation in the scientific sessions and discussions. In addition to the talks and discussions, the program also included a number of social events: a welcome reception at the Toruń branch of the N. Copernicus Astronomical Center, guided tours to the N. Copernicus Family House and St. Johns cathedral church, the concert given by the Hevelius Brass Quintet, and the conference dinner. We also would like to thank the Local Organizing Committee, the staff of the Toruń branch of the N. Copernicus Astronomical Center and graduate students from both the Center and of the Nicolaus Copernicus University. Our particular thanks are due to Ryszard Szczerba, Barbara Gertner and Natasza Siódmiak for taking care of all the events and the excellent local organization. Finally, funds provided in the framework of the SALT network by the Polish Ministry of Scientific Research and Information Technology, which have enabled us to organize the workshop and to support some of the conference participants, are gratefully acknowledged.

Joanna Mikołajewska and Arkadiusz Olech
Warsaw, December 2004

SCIENTIFIC ORGANIZING COMMITTEE

S. Dreizler (Universitaets-Sternwarte Goettingen, Germany)
T. Henning (Max-Planck Institute for Astronomy, Heidelberg, Germany)
J. Kałużny (N. Copernicus Astronomical Center, Warsaw, Poland)
J. Mikołajewska (N. Copernicus Astronomical Center, Warsaw, Poland, co-chair)
B. Paczyński (Princeton University, USA, co-chair)
L. W. Ramsey (HET/Pennsylvania State University, USA)
M. Shara (American Museum of Natural History, USA)
P. A. Whitelock (South African Astronomical Observatory, South Africa)

LOCAL ORGANIZING COMMITTEE

J. Mikołajewska (N. Copernicus Astronomical Center, Warsaw, chair)
R. Moderski (N. Copernicus Astronomical Center, Warsaw)
A. Niedzielski (Centre for Astronomy, N. Copernicus University, Toruń)
R. Szczerba (N. Copernicus Astronomical Center, Toruń)

NICOLAUS COPERNICUS

STELLAR ASTROPHYSICS WITH LARGE TELESCOPES

Stellar Astrophysics on the Hobby-Eberly Telescope [1]

Lawrence W. Ramsey

Department of Astronomy & Astrophysics, The Pennsylvania State University, 525 Davey Lab, Univeristy Park, PA 16802, USA

Abstract.
We briefly describe the Hobby-Eberly Telescope (HET) and its present status and give several examples of successful stellar astrophysics programs. The HET was primarily driven by a science mission of conducting spectroscopic surveys. In addition the HET's unique design derives from considering that telescopes are largely used for exposure times of an hour or less at moderate zenith distances and modest image quality that is based on median site seeing which is consistent with a survey mission. We present results from planet searches, large radial velocity surveys, and Sloan Digital Sky Survey follow up programs in this context. The queue-scheduled nature of the HET makes it especially effective at synoptic programs that require a large aperture.

INTRODUCTION TO THE HOBBY-EBERLY TELESCOPE

The Hobby-Eberly telescope (HET) is an international collaboration between The Pennsylvania State University, The University of Texas at Austin and Stanford University in the United States, and Ludwig-Maxmilians Universität München, and Georg-August-Universität Göttingen in Germany. It is located at the University of Texas' McDonald Observatory at an altitude of 2000 meters. This southwestern site remains exceptional among developed US mainland sites for it dark sky.

The HET is a unique large telescope design making use of an Arecibo-like tracking scheme to allow the implementation of a large aperture telescope for moderate cost by keeping the 11-m spherical primary mirror at a fixed elevation of 35°. A 9.2 meter pupil sweeps over the primary mirror as a x,y tracker moves a spherical aberration corrector to track a target. Objects can be tracked for as long as 2.8 hours at the northern declination limit of $+71.6°$ and for 40 minutes at the southern declination limit of $-10.3°$. A four mirror double-Gregorian corrector design combined with the primary mirror and tracking and alignment errors is designed to yield a 0.6 arcsecond image at the center of a 50 mm (4 arcminute) circular science field of view.

The design, performance goals and early testing is described by Ramsey et al. [1], Sebring et al. [2,3] and Ramsey et al. [4]. Early performance and commissioning difficulties are discussed by Hill et al. [5] and Booth et al. [6] but substantial efforts in the last two years to improve telescope systems and master the very complex operations in-

[1] Based on observations obtained with the Hobby-Eberly Telescope, which is a joint project of the University of Texas at Austin, the Pennsylvania State University, Stanford University, Ludwig-Maxmilians Universität München, and Georg-August-Universität Göttingen.

CP752, *Stellar Astrophysics with the World's Largest Telescopes: First International Workshop*
edited by J. Mikołajewska and A. Olech

herent in this telescope concept have brought the HET to routinely delivering site seeing limited images in the 1.3 arcsecond range. We are confident that further improvement will allow us to exceed our original goals. Current status and performance is presented by Booth et al. [7] and Hill et al. [8]. The HET design has been used in the South African Large Telescope (SALT) where they benefited by our mistakes and we have in turned learned from their improvements (Meiring and Buckley [9]).

FIGURE 1. Hobby-Eberly telescope at McDonald Observatory near Ft. Davis, Texas.

HET INSTRUMENTS

Low resolution spectrograph

The Marcario Low Resolution Spectrograph (LRS) is a Grism based instrument with three modes; broad and narrow band imaging over the HET 4 arcminute field of view, long slit and multi-object spectroscopy (Hill et al. [10,11]). The instrument PI is Dr. Gary Hill and Dr. Phillip MacQueen is PI for the CCD system. The optics were partly designed and constructed in Mexico at IAUNAM (Cobos et al. [12]). This was the first instrument on the HET being commissioned in Spring 1999. It was funded primarily by UT Austin, Stanford University and Ludwig-Maximilians-Universität München. Three grisms are available resolutions of $R = 600(4150 - 10100\ \text{Å})$, $1300(4300 - 7300\ \text{Å})$ and

1900(6250 – 9100 Å). Space constraints on the Prime Focus Instrument Package (PFIP), where the LRS resides, limit only two grisms in the instrument at a time. The position angle of the PFIP can be rotated to set up to arbitrary angles allowing exposures of more than one object simultaneously in the long slit mode and maximizing the number of objects in the remotely configurable 13-slitlet MOS unit.

Medium resolution spectrograph

The Medium Resolution Spectrograph (MRS) is the newest instrument on the HET and is more fully described Ramsey et al. [13,14]. It is a versatile, fiber-fed echelle spectrograph designed for a wide range of scientific investigations and includes single-fiber inputs for the study of point-like sources, synthetic slits of fibers for long slit spectroscopy, 9 independently positionable probes for multi-object spectroscopy, and a circular fiber integral field unit. The MRS consists of two beams. The visible beam has wavelength coverage from 450 - 900 nm in a single exposure with resolving power between 5300 and 20000 depending on the fibers configuration selected. This beam also has capability in the ranges 380 - 950 nm by altering the angles of the cross-disperser gratings. A second beam operating in the near-infrared has coverage of 900 - 1300 nm with resolving power between 5300 and 10000. Both beams can be used simultaneously and are fed by the HET Fiber Instrument Feed (FIF) which is mounted at the prime focus of the telescope and positions the fibers feeding the MRS (Horner et al. [15]). The MRS was funded by Penn State University, Robert E. Eberly and the Eberly family trust and the US National Science Foundation. The PI is L. Ramsey and commissioning began in summer 2003. There are numerous modes on the MRS and at this time, on the simplest modes in the visible beam are used in routine queue operations.

High resolution spectrograph

The High Resolution Spectrograph (HRS) design enables R = 15000 to 120000 spectroscopy on the HET and is described by Tull [16]. It is an adaptation of the white pupil design used on the UVES. It is a fiber coupled, bench mounted instrument with a 420 to 1100 nm wavelength coverage. The HRS is designed for single object use with either 2 or 3 arc-second fibers and a choice of 0, 1, or 2 sky fibers. It uses the same Fiber Instrument Feed as the MRS. The HRS was funded by UT Austin and the US National Science Foundation and commissioned in Spring 2001. The instrument PI was Dr. Bob Tull and Dr. Phillip MacQueen PI for the CCD system. The HRS has proved to be particularly potent for very high precision radial velocity surveys.

SCIENCE WITH THE HET

The HET is a powerful yet limited telescope; a fact that is clear when one looks at the ~\$20 million cost for this 9 meter class telescope. The limitations inherent in the HET design primarily effect scheduling and operational flexibility. Given this the HET will be especially competitive when used with the following criteria in mind:

- Target classes are uniformly distributed on the sky
- Target objects have sky surface densities of a few per square degree or a few per square arcminute
- Time critical observations with time scales of days and longer are of interest
- Spectroscopy in the visible and near infrared yield the required astrophysics

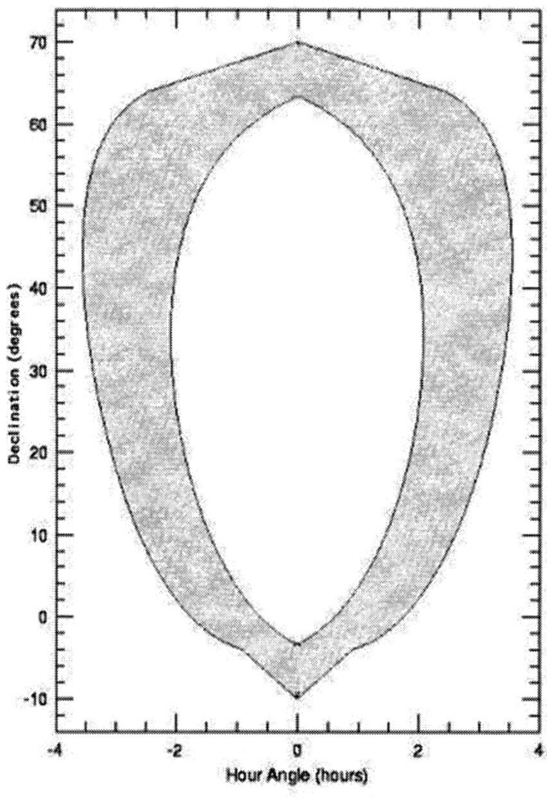

FIGURE 2. Sky access of the HET is illustrated by the shaded area.

The Hobby-Eberly telescope is a 100% queue scheduled telescope. This mode of operation is made necessary by the design of the telescope. Key to understanding this is appreciating the region of the sky that can be accessed at any given time as is illustrated in Figure 2.

All programs on the HET can be classified as survey, synoptic, target of opportunity (ToO), or more traditional programs. Survey programs that are best suited for the HET are those that have large numbers of targets distributed all over the sky or over large areas of the sky. For these programs there are usually several potential targets at any time in the access area shown in Figure 2. Pencil beam surveys, ones which have large number of objects concentrated in a small range of RA and DEC, are less suitable for the HET as any given area can only be accessed twice per night typically for two hours. Studies of the areas such as the Hubble Deep Field or star formation regions typify such programs.

Programs that study the time domain in objects, or synoptic programs are also well suited for the HET especially if the interval of interest is days or longer. These programs require multiple visits to a target or visits on a specific date or dates. While the queue nature of the HET is an asset here, for programs to be successful they have to have high priority relative to all other programs accessible in the same time interval.

Target of opportunity programs are related to synoptic but differ in that they are triggered at some point of time during the observing semester. Programs for Gamma Ray Bursts (GRBs), supernova and nova outbursts typify this type of observation. Of course to be executable at all, the target must be in the sky access ring of the HET as illustrated in Figure 2 at the time of observations.

More traditional programs are those that require long exposure times on a few objects. These will often require multiple tracks and, while they can be done, are not as well suited for the HET. We provide tools on the HET web page so observers can calculate the number of tracks that are likely to be available and estimate the feasibility of their program. Extensive programs in this category must have high priority to have an expectation of being completed.

Successful use of the HET requires that observers must adapt to the nature of queue observing and the HET queue in particular. Observers must make sure that their program is feasible to complete on the HET. At any given location on the sky, there are a limited number of tracking opportunities during an observing trimester. Observers should also honestly assess the competitiveness of their program on HET. During the execution of their programs, observers must promptly look at their data and decide if their program goals are being met. Again the opportunities for any given program diminish with time. This places strong emphasis on communications between the PI and the HET resident astronomers conducting the observations is vital, especially when establishing a new program.

STELLAR ASTOPHYSICS WITH THE HET

As of September 2004 there have been 60 papers published in refereed journal using data from HET. For about 75% of those papers, HET was a significant contributor to the paper. In the early work (2000-2002) HET publications benefited greatly form working with SDSS collaborators. Papers that are based nearly 100% on HET data are common recently. Of the total number of publications 37%, or 22 papers, are stellar astrophysics.

Surveys

To date, survey science, along with synoptic science, has produced the largest number of stellar astrophysics papers with 7 publications. The first stellar astrophysics done on the HET was limited to using the LRS and was mostly due to productive collaborations with the Sloan Digital Sky Survey teams working on M, L and T dwarf stars (Leggett et al. [17], Schneider et al. [18], Geballe et al. [19].). An early LRS survey also studied the spectral properties of faint UV bright stars found on object prism surveys (Eracleous et al. [20]) and more recent LRS observations have been used to complement photometric identifications of cool white dwarfs (Kilic et al. [21]). Yong and Lambert [22] used an earlier medium-resolution commissioning instrument to survey 134 candidates for cool subdwarfs. Allende-Prieto et al [23] conducted the first survey with the HRS using $R = 120000$ high S/N spectra to look at the convective shifts of over 1200 FeI lines to show clear evidence of a greater shift in weaker lines and an increase in amplitude for more luminous stars.

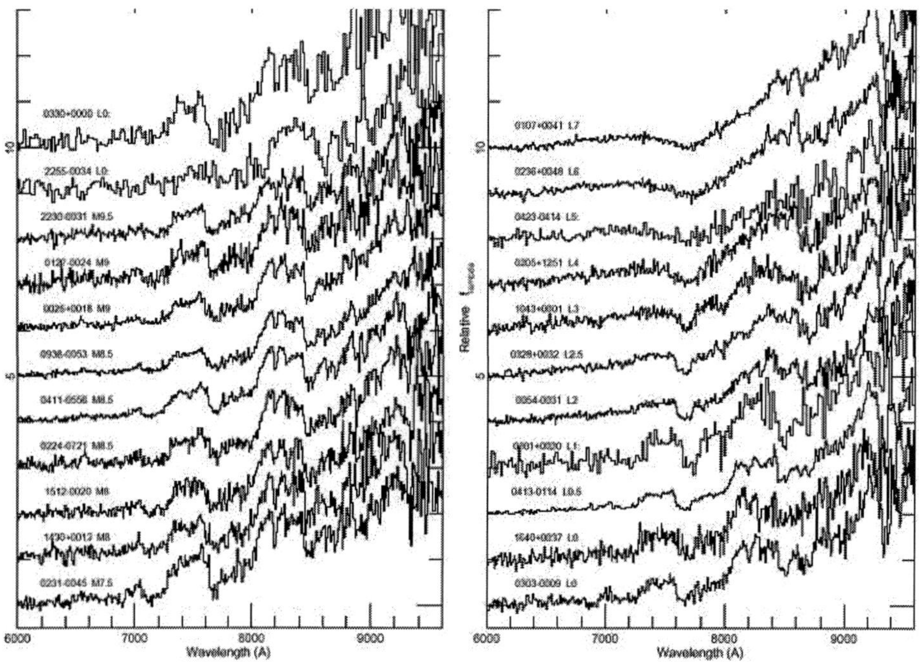

FIGURE 3. HET spectra of M, L and T dwarfs obtained using the LRS.

One of the early survey results was that of Schneider et al. [18] which showed that SDSS colors allow identification of L and T dwarfs with a very high success rate. They also showed that a combination of factors, the disappearance of TiO and VO in the i band for latter types, the presence of strong NaI and KI in the i band and the disappearance of 840 nm TiO and FeH in the z band, made the SDSS $i-z$ color insensitive to spectral type from M7 to L4.

Synoptic Science

Synoptic science comprises most of the recent papers and has the highest proportion of papers that depend mostly on HET data. This is not surprising given the high effectiveness the HRS on HET has for Radial Velocity surveys which account for 5 of the 8 papers in this category to date (Endl et al. [24], Paulson et al. [25,26], Cochran et al. [27] and McAuthur et al. [28]). Indeed all the publications in this category have depend upon HRS data. Smaller synoptic programs have looked at stellar activity (Lopez-Santiago et al. [29]), variability in the NaD line to find evidence of comet infall (Mendelowitz [30]) and transient features in the sodium lines in young stars for evidence of planetesimal infall (Chakraborty [31]). Likely the most important paper in the synoptic category is that by McArthur et al. [28] where they obtain lovely data showing the existence of a Neptune mass planet in the ρ^1 Cancri system. The high precision of the HRS and the flexible queue schedule of the HET made this and the discovery of another planet (Cochran et al. [27]) possible over a very short time scale. Figure 4 is a summary of recent HRS measurements compiled by Dr. William Cochran and his collaborators in the HET planet search program where measurements less than 5 km/sec are routine.

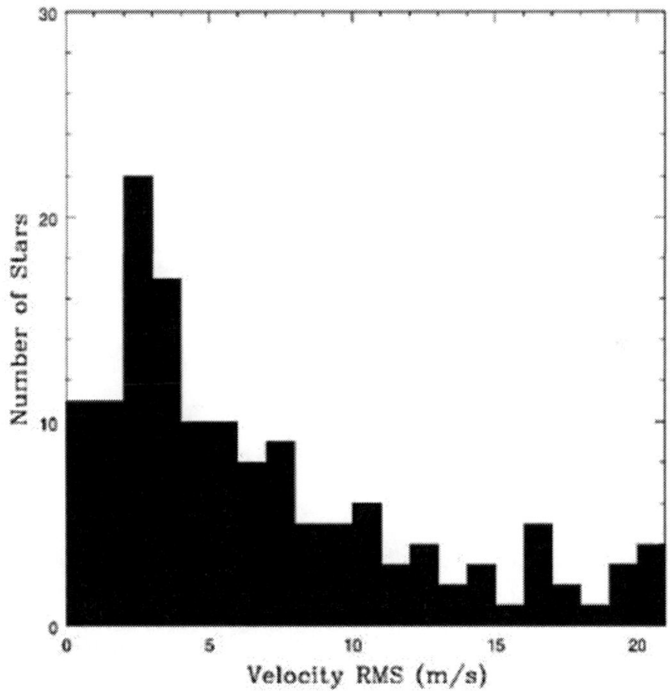

FIGURE 4. Histrogram showing the radial velocity precision of the HET for a survey of F and K stars with V magnitudes ranging from 6 to 10.

Target of Opportunity Science

Target of opportunity science has not yet hit stride with the HET. While such programs should be well suited for HET, they have lagged primarily because of the lack of robustness in our operation until this last year. In spite of these difficulties, some results have been published. Indeed one of the first papers from HET used the LRS to observe a 4.2 hour eclipsing dwarf nova (Gansicke et al. [32]). More interesting results were obtained on SN 2003du where an nice LRS time series showed the evolution of the Ca II infrared triplet thought to arise in the circumstellar environment (Gerardy et al. [33]). We have had a limited success with GRB's (Schaefer et al. [34]) but hope to do better in this area in the future.

Traditional Stellar Astrophysics

While more traditional programs are not HET strength, thoughtful and patient work can pay off. Matthew Shetrone, who is the lead astronomer at HET, and collaborators has published several papers probing aspects of stellar evolution [35,36,37]. Also, a very nice "traditional" study was carried out by Mandell et al. [38] where they used $R = 120000$ very high S/N spectra of the Li line region in Li poor stars ^{6}Li that would be evidence for accretion of planets.

FUTURE PROSPECTS

As the performance of the HET continues to improve we anticipate an increase in the productivity of the telescope and its instruments. The MRS has been in the queue observing mode for less than a year and the number of programs using it is increasing. In addition, both the LRS and the MRS will have J band (0.9-1.3 micron) capability within a year. We expect that most stellar astrophysics will continue to be done with the HRS followed by the MRS.

ACKNOWLEDGMENTS

The success of the HET is due to a large number of people. In addition to the authors and co-authors cited above I would like to acknowledge the many contributions of Gary Hill, Phillip MacQueen, Polilas Palunas, Marsha Wolf. I also appreciate the continued hard work by the HET support team lead by John Booth which includes Gordon Wesley, Edmundo Balderamma, Linda Brindle, George Damm, Fernando De Valesco, Jim Fowler, Paul Peterson, Robert Poenisch and Mike Ward, HET resident astronomers Steve Odewahn, Brian Roman and Matthew Shetrone, telescope operators, Frank Deglman, Vicky Riley and Martin Villareal.

REFERENCES

1. Ramsey, L.W., Sebring, T.A., and Sneden, C., "The Spectroscopic Survey Telescope Project", Proceedings of the SPIE, **2199**, Advanced Technology Optical Telescopes V, p. 31 (1994).
2. Sebring, T.A., Booth, J.A., Good, J.M., Krabbendam, V.L., and Ray, F.B., "Design and Status of the Spectroscopic Survey Telescope", Proceedings of the SPIE, **2199**, Advanced Technology Optical Telescopes V, p. 565 (1994).
3. Sebring, T.A. and Ramsey, L.W. "The Hobby-Eberly Telescope: A progress report", Proceedings of the SPIE, **2871**, Optical Telescopes of Today and Tomorrow, p. 32. (1997).
4. L.W. Ramsey, M.T. Adams, T.G. Barnes, J.A. Booth, M.E. Cornell, J.R. Fowler, N.I. Gaffney, J.W. Glaspey, J. Good, P.W. Kelton, V.L. Krabbendam, L. Long, F.B. Ray, R.L. Ricklefs, J. Sage, T.A. Sebring, W.J. Spiesman, and M. Steiner, "The early performance and present status of the Hobby-Eberly Telescope," Proceedings of the SPIE, **3352**, Advanced Technology Optical/IR Telescopes VI, p. 34 (1998).
5. G.H. Hill, P.J. MacQueen, L.W. Ramsey, "Performance of the facility instruments on the Hobby-Eberly telescope", Proceedings of the SPIE, **4841**, p. 43, (2003)
6. Booth, J.A. et al., "The Hobby-Eberly Telescope Completion Project" in *Large Ground-based Telescopes.*, edited by J.M. Oschmann and L.M Stepp, Proceedings of the SPIE, **4837**, pp. 919-933 (2003).
7. Booth, J.A. et al., "The Hobby-Eberly Telescope: performance upgrades, status, and plans", Proceedings of the SPIE, **5489**, pp. 288-299 (2004).
8. G.H. Hill, P.J. MacQueen, L.W. Ramsey and M.D. Shetrone "Performance of the Hobby-Eberly telescope facility instruments on", Proceedings of the SPIE, **5492**, pp. 94-104 (2004).
9. Mering, K. and Buckley, D.H. "Southern African Large Telescope (SALT) project progress and status after four years", Proceedings of the SPIE, **5489**, pp. 592-602 (2004).
10. Hill, G.J., Nicklas, H., MacQueen, P.J., Tejada , C., Cobos, F.J., and Mitsch, W., "The Hobby-Eberly telescope Low Resolution Spectrograph", Proceedings of the SPIE, **3352**, Optical Astronomical Instrumentation, p. 375 (1998).
11. Hill, G.J., Nicklas, H.E., MacQueen, P.J., Tejada, C., Cobos D., Francisco J., Mitsch, W., "Hobby-Eberly Telescope low-resolution spectrograph", Procceedings of the SPIE, **3355**, Optical Astronomical Instrumentation, pp. 375-386 (1998).
12. Cobos D., Francisco J., Tejada, C., Hill, G.J., Perez G.F., "Hobby-Eberly Telescope low-resolution spectrograph: optical design", Proceedings of the SPIE, **3355**, Optical Astronomical Instrumentation, pp. 424-432 (1998).
13. Ramsey, L.W. "Hobby-Eberly telescope medium-resolution spectrograph", Proceedings of the SPIE, **2476**, Fiber Optics in Astronomical Applications, pp. 20-32 (1995).
14. Ramsey, L.W., Engel, L.G., Sessions, N., DeFilippo, C., Graver, M. and Mader, J., "The Hobby-Eberly Telescope medium resolution spectrograph and fiber instrument feed" in *Instrument Design and Performance for Optical/Infrared Ground-based Telescopes*, edited by Iye, M. and Moorwood, A.F.M., Proceedings of the SPIE, **4841**, pp. 1036-1044 (2003).
15. S. D. Horner, L. G. Engel, L. W. Ramsey, "Hobby Eberly Telescope medium-resolution spectrograph and fiber instrument feed", Proceedings of the SPIE, **3355**, Part One, Optical Astronomical Instrumentation, p. 399 (1998).
16. Tull, R.G., "High resolution fiber coupled spectrograph of the Hobby-Eberly Telescope", Proceedings of the SPIE **3355**, Optical Astronomical Instrumentation, pp. 387-398 (1998).
17. S.K. Leggett, T.R. Geballe, X. Fan, D.P. Schneider, J.E. Gunn, R.H. Lupton, G.R. Knapp, M.A. Strauss, A. McDaniel, D. Golimowski, T. Henry, E. Peng, Z.I. Tsvetanov, A. Uomoto, W. Zheng, G.J. Hill, L.W. Ramsey, S.F. Anderson, N.A. Bahcall, J. Brinkmann, B. Chen, I. Csabai, M. Fukugita, G.S. Hennessy, R.B. Hindsley, Z. Ivezic, D.Q. Lamb, J.A. Munn, J.R. Pier, J.A. Smith, C. Stoughton, A.R. Thakar, and D.G. York, *ApJ Letters*, **536**, L35-L38 (2000).
18. Schneider, D.P., Knapp, G.R., Hawley, S.L. et al., *AJ*, **123**, 458-465 (2002).
19. Geballe, T.R., Knapp, G.R., Leggett, S.K., Fan, X., Golimowski, D, Tsvetanov, Z.I., Gunn, J.E., Lupton, R.H., Strauss, M.A., McDaniel, A., Narayanan, V., Henery, T., Peng, E., Uomoto, A., Zheng, W., Schneider, D.P., Hill, G.J., Brinkman, J., *ApJ*, **564**, 466-481 (2002).
20. Eracleous, M., Wade, R.A., Mateen M., and Lanning, H., *PASP*, **114**, 207-213 (2002).
21. Kilic, M., Winget, D.E., von Hippel T., and Claver, C.F. *AJ*, **128**, 1825-1833 (2004).

22. Yong, D. and Lambert, D.L., *PASP*, **115**, 22-36 (2003).
23. Allende, P.C., Lambert, D.L., Tull, R.G., MacQueen, P.J., *ApJ Letters*, **566**, L93-L96 (2002).
24. Endl, M., Cochran, W.D., Tull, R.G., MacQueen, P.J., *AJ*, **126**, 3099-3107 (2003).
25. D. Paulson, S. Saar, W. Cochran, G. Henry, *AJ*, **127**, 1644 (2004).
26. Paulson,ăD.B., Cochran,ăW.D., Hatzes,ăA.P., *AJ*, **127**, 3579-3586 (2004).
27. Cochran, W.D., Endl, M., McArthur, B. et al., *ApJ Letters*, **611**, L133-L136 (2004).
28. McArthur, B.E., Endl, M., Cochran, W.D. et al. *ApJ*, **614**, L81-L84
29. Lopez-Santiago, J., Montes, D., Fernandez-Figueroa, M.J., Ramsey, L.W., *A&A*, **411**, 489-502 (2003).
30. Mendelowitz, C., Ge, J., Mandell, A.M., Li, A., *ApJ*, **601**, 1120-1128 (2004).
31. Chakraborty, A., Ge, J., Mahadevan S., *ApJ Letters*, **606**, L69-L72 (2004).
32. Gänsicke, B.T., Fried, R.E., Hagen, H.-J., Beuermann, K., Engels, D., Hessman, F.V., Nogami, D., Reinsch, K., *A&A* **356**, L79-L82 (2000).
33. Gerardy C., Hoeflich, P., Fesen, R.A., Marion, G.H., Nomoto, K., Quimby, R., Schaefer, B.E., Wang, L., Wheeler, C.J., *ApJ*, **607**, 391 (2004).
34. Schaefer, B.E., Gerardy, C.L., Hoflich, P. et al., *ApJ*, **588**, 387-399 (2003)
35. Shetrone, M.D., Sandquist, E.L., *AJ*, **120**, 1913-1924 (2000).
36. Shetrone, M.D., Cote, P., Stetson, P.B., *PASP*, **113**, 1122-1129 (2001).
37. Sandquist, E.L., Latham, D.W., Shetrone, M.D., Milone, A.A.E., *AJ*, **125** 810-824 (2003).
38. Mandell, A.M., Ge, J., Murray, N., *AJ*, **127**, 1147-1157 (2004).

Stellar atmosphere modeling in the era of 10m class telescopes

S. Dreizler

Institut für Astrophysik, Friedrich-Hund-Platz 1, 37077 Göttingen, Germany

Abstract. In this paper, I briefly summarize the basic ingredients for modeling stellar atmospheres. The progress in the numerical solution techniques has evolved in parallel to the rapid development of observing facilities. The modeling of stellar atmospheres in particular and stellar astrophysics in general are therefore important ingredients for our understanding of stellar, galactic and cosmic evolution.

INTRODUCTION

Photons are the most important source of information for the investigation of astrophysical objects and most of the photons are produced by stars or similarly optically thick objects. Basically, all our knowledge about structure and evolution of stars rests on the interpretation of the electromagnetic spectrum. The spectral characteristics reveal temperatures, densities, chemical compositions as well as magnetic fields or velocity fields. The temporal variation allows to explore dynamical and structural properties, e.g. masses of objects in multiple systems, surface inhomogeneities on rotating stars or the stellar structure determinations from pulsating stars. A sound knowledge about stellar astrophysics is also a prerequisite for other fields, like galactic and cosmic evolution. The quantitative analysis of stellar spectra is therefore an important tool of modern astrophysics.

STELLAR ATMOSPHERES

The formation of an observed spectrum is usually confined to a very thin layer on top of the stellar core, the atmosphere. In this transition zone, the radiation transport has to be modeled in detail in order to correctly reproduce the spectral characteristics. The physical information is then extracted by comparing theoretical spectra with observations. The physical assumptions under which the radiation transport and therefore theoretical spectra are usually solved are summarized in the following. Due to the rapid development of the field during the last decade, a complete overview is far beyond the scope of this paper. A rather personally biased view will therefore be given here. For a more detailed description we refer to e.g. Hauschildt & Baron [1] and Werner & Dreizler [2] or to the text book of Mihalas [3].

CP752, *Stellar Astrophysics with the World's Largest Telescopes: First International Workshop*
edited by J. Mikołajewska and A. Olech

Radiation transfer

The basic physical quantity is the radiation field I_ν which is, except for constant factors, the pointing vector of the electromagnetic field at the frequency ν. The calculation of a theoretical spectrum as well as the knowledge of the radiation field at any point in the stellar atmosphere requires the solution of the radiation transport equation along a path element s:

$$\frac{dI_\nu}{ds} = -\kappa(\nu)I_\nu + \eta_\nu. \tag{1}$$

The first term on the right-hand side describes the losses, the second the gains due to absorption (κ) and emission (η) processes. In most applications, this equation can be solved assuming plane-parallel geometry. The justification for this assumption is the fact that the extension of a typical stellar atmosphere is orders magnitude smaller than the stellar radius and curvature effects are therefore negligible. Exceptions are for example atmospheres of giants as well as expanding envelopes, which do require spherical symmetry. While 2-D models are also possible with current solution techniques, the full 3-D problem with detailed atomic physics is still beyond current computer capacities.

The differential equation above can be solved in case the absorption and emission processes are given. This requires the knowledge of all important interaction processes of photons with electrons, atoms, ions and/or molecules. While the necessary cross sections are basically physical constants, the absorption and emission also depends on the population numbers of the energy levels of the absorbing/emitting atom, ion or molecule. We therefore have to know the temperature and density as well as the populating and de-populating processes for the energy levels, which in turn requires the knowledge of the radiation field (equation 1) at every point in the atmosphere.

Density stratification - hydrostatic equilibrium

The densities can be derived from the assumption of hydrostatic equilibrium, which is valid for the vast majority of stars. The following equation is written for a plane-parallel depth grid (μ being the cosine of the angle between the ray and the vertical direction) on a column mass density m and involves the gas and radiation pressure (left-hand side) which are balanced by the gravitational acceleration g:

$$\frac{d}{dm}\left(NkT + \frac{2\pi}{c}\int_0^\infty \int_{-1}^{1} \mu^2 I_\nu(\mu)\, d\mu\, d\nu\right) = g. \tag{2}$$

Additional terms like turbulent or magnetic pressure can also be added. The assumption of hydrostatic equilibrium has to be replaced by a continuity equation in case of expanding atmospheres, e.g. when the radiative acceleration exceeds the gravitational acceleration. It should be noted, that the solution requires the knowledge of the radiation field at any depth point from equation (1).

Temperature stratification - radiative equilibrium

In general, the temperature has to decrease from the stellar interior towards the outside, since this temperature gradient is responsible for the energy transport from the interior, where the energy is generated from nuclear processes, towards the stellar surface, where the energy is lost into the interstellar medium. Energy transport is either via photons or via convection, depending on the temperature gradient. Typically, no energy is generated within the atmosphere, for the radiative equilibrium case a net balance between absorbed and re-emitted radiation energy is required to cancel:

$$\int_0^{\infty} \left(\eta_\nu - \int_{-1}^{1} \kappa(\nu) I_\nu d\mu \right) d\nu = 0. \tag{3}$$

The contributions of convective energy transport can be handled with a non-vanishing right-hand side. A non-vanishing right-hand side can also be used to describe energy release e.g. by radiative decays of radioactive species during a supernova explosion. Again, the radiation field at any depth point from equation (1) is required.

In case the additional energy release is unknown or difficult to parameterize (e.g. in the outer atmosphere of the sun), the temperature stratification has to be treated as a free function which must be adjusted by the comparison of observation and model.

Population densities - statistical equilibrium

A stellar atmosphere radiates into interstellar space and as an open thermodynamic system hence cannot be in thermodynamic equilibrium (TE): We cannot simply assign temperatures. "Local Thermodynamic Equilibrium" (LTE) is a working hypothesis which assumes TE exists not for the atmosphere as a whole but for small volume elements. As a consequence, the atomic population numbers depend only on the local (electron) temperature and electron density via the Saha-Boltzmann equations. This assumption is valid as long as the electron collisions dominate the population and depopulation processes, i.e. electron collisions are more frequent than radiative processes. Higher densities and weak radiation fields therefore favor the LTE assumption. Since the density decreases outwards, the LTE approximation finally fails in every star. With increasing temperatures the radiation field increases with the fourth power, hotter stars are therefore less suitable for the LTE approximation. The limits of assuming LTE depend on the degree of sophistication, both of the observation as well as of the analysis. Relaxing the LTE assumption requires an replacement of the Saha-Boltzmann equations. Under so-called non-LTE conditions, all populating and de-populating processes have to be accounted in detail. Since the time scales for photon transitions are many orders of magnitude shorter than the time scales for changes of the physical conditions in stellar atmospheres, e.g. due to stellar evolution but also for stellar pulsation, an statistical equilibrium can be assumed, i.e. we have a vanishing net change of the population density for all energy levels.

$$n_i \sum_{i \neq j} P_{ij} - \sum_{j \neq i} n_j P_{ji} = \frac{dn_i}{dt} = 0. \tag{4}$$

Such an equation for the population number of all energy levels n_i describes the transitions to all other levels j; the rate coefficients P_{ij} have radiative and collisional components. Densities, temperatures and the radiation field has to known in order to solve the equations.

Solution strategies

All the equations are coupled over all depths of the atmosphere and over all frequencies and are mutually dependent. The numerical effort for a simultaneous solution of all equations is therefore tremendously large; a first attempt to solve the whole non-LTE problem was made in 1969 by Auer & Mihalas [4], but the complexity of the system of equations enforced a restriction to simple atomic models. Significant progress was possible with a de-coupling of the equations. A major step was the introduction of the Accelerated Lambda Iteration (ALI, or "operator splitting" methods), based upon early work by Cannon [5] and Scharmer [6]. This allows to separate the radiation transport from the other equations, the main computational effort can therefore be concentrated in the solution of the equation of statistical equilibrium. With the largely increased numerical capacity the number of energy levels to describe the atomic physics can be increased accordingly. The first ALI model atmospheres were constructed by Werner [7]. Another important step was the introduction of statistical approaches to treat the iron group elements in non-LTE by Anderson [8, 9], allowing the treatment of millions of spectral lines under non-LTE conditions. A final step is the separation of the energy equation by the introduction of temperature correction methods (Hauschildt et al., Dreizler, [10, 11]). The last step allows the solution of the equations for radiation transport, hydrostatic equilibrium, radiative equilibrium, and statistical equilibrium separated for each atom consecutively, again reducing the computational demand drastically.

APPLICATIONS AND SUMMARY

Applications of stellar atmospheres are manyfold. The most direct application is the determination of stellar parameters which can be coupled with stellar evolution models in order to explore to cosmic circuit of matter from star formation, through stellar evolution and interaction with the interstellar medium during the final stages. With 10m class telescopes such investigations are now not only possible in our Galaxy but can be extended to the local group. A sound understanding of stars also enables us to use them as laboratories for fundamental physics like atoms in high magnetic fields. Alternatively we can use stars as standard candles exploring the interstellar medium or to derive cosmological distances.

The development in numerical methods, the largely improved atomic and molecular data required as input physics together with the increase of computational power here resulted in a huge progress in modeling stellar atmospheres. State-of-the-art models are capable of treating millions of spectral lines also under non-LTE conditions for nearly all physical conditions, e.g. for extremely high or very low temperatures or for extreme densities.

With the continuous progress on the observational side, current and future challenges for stellar atmospheres and radiation transport are adequate models for time resolved spectroscopy (e.g. of pulsating stars), accretion processes (e.g. in close binary systems with accretion streams and disks as well as mutual irradiation), explosive processes (e.g. non-monotonic velocity fields), or atmospheres of highly magnetic objects (e.g. neutron stars).

REFERENCES

1. Hauschildt, P.H., Baron, E. 1999, in The Journal of Computational and Applied Mathematics, Volume 109, eds. H. Riffert & K. Werner, Elsevier Press, Amsterdam, p. 41
2. Werner, K., Dreizler, S. 1999, in The Journal of Computational and Applied Mathematics, Volume 109, eds. H. Riffert & K. Werner, Elsevier Press, Amsterdam, p. 65
3. Mihalas, D. 1978, Stellar Atmospheres, Freeman, San Francisco
4. Auer L.H., Mihalas D. 1969, ApJ 158, 641
5. Cannon C.J. 1973, ApJ 185, 621
6. Scharmer G.B. 1981, ApJ 249, 720
7. Werner K. 1986, A&A 161, 177
8. Anderson L.S. 1989, ApJ 339, 558
9. Anderson L.S. 1991, in Stellar Atmospheres: Beyond Classical Models, NATO ASI Series C, Vol. 341, eds. L.Crivellari, I.Hubeny, D.G.Hummer, Kluwer, Dordrecht, p. 29
10. Hauschildt, P.H., Barman, T.S., Baron, E., Allard, F. 2003, ASP Conf. Ser. 288, eds. I. Hubeny, D. Mihalas & K. Werner: Stellar Atmosphere Modeling, 227
11. Dreizler, S. 2003, ASP Conf. Ser. 288, eds. I. Hubeny, D. Mihalas & K. Werner: Stellar Atmosphere Modeling, 69

The SALT First Generation Instruments: Ideal Tools for Stellar Astrophysics

David A. H. Buckley

Southern African Large Telescope
South African Astronomical Observatory
PO Box 9, Observatory 7935, Cape Town, South Africa

Abstract. SALT's first generation instrumentation will be ideally suited to stellar astrophysical studies, both in our own Galaxy and in nearby galaxies. This paper will describe the design, capabilities and status of the instruments, emphasizing those modes and niches that will impact stellar astronomy. The first-light 'visible beam' (320-900nm) instruments, the imaging camera, SALTICAM, and the imaging spectrograph, PFIS, are due to begin commissioning observations later in 2004/early 2005. The latter will be the work-horse instrument on SALT in the near future, and contains many modes appropriate for stellar observational astronomy, namely low to medium resolution spectroscopy and spectropolarimetry, Fabry-Perot imaging spectroscopy, narrow band imaging, and imaging polarimetry. Both instruments will support high time resolution observations, one of the niche areas of stellar astrophysics. The last of the first generation instruments will be a fibre-fed double-beam high resolution white-pupil R4 échelle spectrograph, SALT HRS, designed for good stability, a range or resolutions and accurate background subtraction. The instrument is currently in its design phase, and construction will likely begin in early 2005, with an estimated delivery in 2007.

INTRODUCTION

The Southern African Large Telescope (SALT) is a 10-m class optical/IR telescope [1], based on the innovative, low cost, Hobby-Eberly Telescope (HET) 'tilted Arecibo' design [2]. When completed in mid-2005, SALT, which is situated at the Sutherland observing station of the South African Astronomical Observatory, will be the largest single optical telescope in the Southern Hemisphere. Approval to begin construction was given by the South African Government in November 1999, after matching funding from SALT's international partners[1] were secured. Design work then began and construction started in early 2001.

SALT will be operated entirely as a queue scheduled telescope, able to access targets in the declination range $-75° \leq \delta \leq +10°$ and observe them continuously during the night for periods of 1 – 3 hours, depending on δ, or even longer in certain circumstances (e.g. allowing azimuth shifts of the telescope structure).

[1] The SALT Foundation currently comprises partners from South Africa, Poland, UK, USA, Germany and New Zealand and includes the National Research Foundation of South Africa, the Nicholas Copernicus Astronomy Center, UK SALT Consortium, University of Wisconsin-Madison, Rutgers University, Carnegie Mellon University, University of North Carolina, Hobby Eberly Telescope Board, Dartmouth College, Georg-Augustus University, Göttingen, and the University of Canterbury.

CP752, *Stellar Astrophysics with the World's Largest Telescopes: First International Workshop*
edited by J. Mikołajewska and A. Olech

FIRST-GENERATION INSTRUMENTS

Provision has been made within the SALT construction budget, through a combination of both cash and "in-kind" funding, for a set of three First-Generation science instruments [3]. All three instruments are being built by SALT consortium members and are designed to be seeing-limited (median zenith seeing in the V-band is 0.9 arcsec FWHM) and will take advantage of the expected telescope image quality (~0.54 arcsec FWHM), science field of view (8 arcmin diameter) and good UV/blue response (down to $\lambda = 320$ nm).

SALT's suite of First-Generation instruments [3] will be well suited to stellar astrophysical studies, initially confined to just the visible-UV region (320 – 900 nm), but with planned extensions into the near IR (J and H at least). Two of the instruments (SALTICAM and PFIS, due for completion in February and March 2005, respectively) will be mounted at SALT's prime focus, while the third (HRS, still in the design phase) will be installed in the spectrograph room beneath the telescope, and fed by optical fibers. The characteristics, capabilities and status of these instruments are described in the following sections, with details given in [3].

PFIS: the Prime Focus Imaging Spectrograph

PFIS will be the work-horse instrument on SALT, and is currently in construction at the University of Wisconsin-Madison, with installation scheduled for March 2005 [4]. Two other SALT partners are also involved in PFIS: Rutgers University are responsible for the Fabry-Perot etalons and the Invar structure, while the South African Astronomical Observatory (SAAO) has built the CCD detector, consisting of a mosaic of thee 2K × 4K CCDs .

PFIS has been designed with a range of capabilities and modes [4, 5], each one remotely and rapidly able to be reconfigured. In keeping with the overall philosophy of exploiting the niche areas where SALT has a competitive edge, the instrument has several unique or rare capabilities, some afforded by various enabling technologies, which include:

- Observing to the UV atmospheric cut-off, at ~320 nm. This is achieved using UV transmitting materials and efficient anti-reflection coatings on the optics.
- A fully articulating camera/detector used with a suite of five efficient Volume Phase Holographic transmission gratings (VPHGs) and a surface relief grating.
- All-Stokes mode polarimetric capability [6], using both ½ and ¼ waveplate retarders and a large mosaiced Wollaston beam-splitter, which completely separates the two O- and E-images on the CCD detector.
- Fabry-Perot imaging spectroscopy from 430-860 nm using three etalons, with up to two in series, giving resolutions of R = 500-1000, ~2500 and ~12,500.
- The use of fast frame-transfer CCDs allowing for high-speed observations in all observing modes.

These capabilities will give PFIS a wavelength coverage from ~320 nm to 900 nm, and a near-IR (to 1.7 microns) upgrade path using a dichroic beamsplitter. PFIS will undertake low to medium resolution spectroscopy (up to R ~ 6000 with 1 arcsec slits) using up to five efficient and tuneable VPH gratings and one surface relief grating. Multiple object spectroscopy (MOS), of up to ~100 objects at a time, will be possible using laser cut graphite focal plane slit masks. In addition, narrow to intermediate band imaging will be possible.

The CCDs for PFIS, and SALTICAM, are E2V CCD 44-82 chips, each with 2K × 4K × 15μm pixels, and in a mosaic of 3 and 2, respectively. These CCDs are thinned, back-illuminated deep depletion devices, coated for good UV response. They are designed to run in frame transfer mode for high time resolution, which should allow for observations at ~10Hz for either high-speed spectroscopy or windowed differential photometry, with insignificant readout time losses. All the chips are cosmetic Grade 0 devices with low readout noise (2.0-2.4ē) and excellent QE (one chip has a QE of 80% at 350 nm).

SALTICAM: SALT Imaging Camera

SALTICAM was conceived to be a multi-purpose instrument, capable of performing roles as both an efficient acquisition camera for SALT, and a high-speed CCD scientific imager [7]. SALTICAM is being built at the SAAO in two stages, the first without focal reduction optics, where it serves the purpose of the SALT Verification and Commissioning Instrument (VI), which was delivered in September 2003. In January 2005, SALTICAM will undergo a transformation from this VI mode to the so-called Acquisition Camera and Science Image (ACSI) mode, when focal reducing fore-optics will be integrated into the instrument. These fore-optics will provide focal reduction (from f/4.2 to f/2) to give a plate scale which allows for the entire science and guidance fields to be imaged onto the 4K × 4K detector area (i.e. a field of ~10 × 10 arcmin).

The same frame transfer, science Grade 0 CCDs, as used in PFIS, will allow for fast acquisition, science imaging and high-time resolution photometry, likely in excess of ~10 Hz using a slot in the moveable frame transfer mask. Together with the PFIS, SALTICAM will expand the parameter domain of time variability studies on 10-m class telescopes. This will have a major impact in certain fields of stellar astronomy, e.g. accretion driven binary stars and rapid variability. Like PFIS, the SALTICAM optics are designed to allow observations down to the UV atmospheric cut off.

HRS: High Resolution Spectrograph

The third First-Generation instrument is the SALT High Resolution Spectrograph (HRS), which is a fibre-fed R4 dual beam white pupil échelle spectrograph, to be built by the University of Canterbury, and is currently in the critical design phase, with commissioning expected to begin late-2006. The concept for this instrument is for an efficient and stable single-object spectrograph using one of five pairs of optical fibers (from 300μm to 500μm in diameter), one for source (star) and one for background (sky). Some of these will feed image slicers, allowing for up to 4 slices in the spatial direction, before injection into a dual-beam white-pupil R4 spectrograph, which will deliver a resolving power of between R =16,000 and 70,000. The wavelength coverage of the two beams, split by a dichroic, will be in the region 370 – 560 nm (blue) and 560 – 870 nm (red). A single 2K × 4K CCD will be sufficient to capture all the blue orders, while a 4K × 4K detector (possibly a mosaic), will be required for the red.

The spectrograph will be enclosed in a vacuum tank to improve stability of the instrument, and to ensure the optics remain in a pristine condition. In addition, to allow for precise measurement of radial velocities, various additional techniques may be applied (e.g. use of an iodine cell, dispersive interferometry, simultaneous injection of ThAr arc spectra). To ensure accurate background subtraction on fainter objects, both beam switching and "nod & shuffle" techniques will be supported by HRS and the Fiber Instrument Feed (FIF), mounted on the prime focus payload, which delivers the light to it.

SALT HRS will be well suited to a range of stellar observational programs, and with SALT's queue scheduling ability, will be particularly well suited to synoptic variability studies, precise radial velocity determinations and searches for exo-planets.

ACKNOWLEDGMENTS

I wish to thank the instrument PIs, Ken Nordsieck, Darragh O'Donoghue and Peter Cottrell, for their commitment to seeing that SALT has excellent instruments which will address many of the science drivers from the SALT community. I am equally indepted to Kobus Meiring, and all of the SALT Project Team, who have made SALT a reality within 5 years – a remarkable effort.

REFERENCES

1 J. G. Meiring, D. A. H. Buckley, M. Lomberg and R. S. Stobie, "The SALT project: progress and status after 2 years", in *Large Ground-based Telescopes*, J.M. Oschmann and L.M. Stepp, eds., *Proc. SPIE* **4837**, pp. 11-25, 2003.

2 L. W. Ramsey, et al., "The early performance and present status of the Hobby-Eberly telescope", in *Advanced Technology Optical/IR Telescopes VI*, L. M. Stepp, ed., *Proc. SPIE* **3352**, pp. 34-42, 1998.

3 D. A. H. Buckley, J. B. Hearnshaw, K. H. Nordsieck and D. O'Donoghue, "Science drivers and first generation instrumentation for SALT", in *Discoveries and Research Prospects from 6- to 10-meter class Telescopes II,* P. Guhathakurta, ed., *Proc. SPIE* **4834**, pp. 264-275.

4 E. Burgh, et al., "The Prime Focus Imaging Spectrograph for the Southern African Large Telescope: optical design", in *Instrument Design and Performance for Optical/Infrared Ground-based Telescopes*, M. Iye and A.F. Moorwood, eds., *Proc. SPIE* **4841**, pp. 1463-1471.

5 H. A. Kobulnicky, et al., "The Prime Focus Imaging Spectrograph for the Southern African Large Telescope: operational modes", in *Instrument Design and Performance for Optical/Infrared Ground-based Telescopes*, M. Iye and A.F. Moorwood, eds., *Proc. SPIE* **4841**, pp. 1634-1644.

6 K. H. Nordsieck, K. P. Jaehnig, E. B. Burgh, H. A. Kobulnicky, J. W. Percival, and M. P. Smith, "Instrumentation for high-resolution spectropolarimetry in the visible and far-ultraviolet", in *Polarimetry in Astronomy*, S. Fineschi, ed., *Proc. SPIE* **4843**, pp. 170-179.

7 D. O'Donoghue, et al., "SALTICAM: a $0.5M acquisition Camera: Every Big Telescope Should Have One", in *Instrument Design and Performance for Optical/Infrared Ground-based Telescopes*, M. Iye and A.F. Moorwood, eds., *Proc. SPIE* **4841**, pp. 465-476.

High-Resolution Spectroscopy in the SALT Era

W. Tobin*, P.L. Cottrell*,†, S.I. Barnes*, M.D. Albrow* and G.M. Kershaw*

*Department of Physics & Astronomy and Mount John University Observatory, University of Canterbury, Private Bag 4800, Christchurch, New Zealand.
†SALT HRS Principal Investigator

Abstract. As part of its contribution to the SALT partnership, the University of Canterbury expects to build a fibre-fed, high-resolution stellar spectrograph with excellent wavelength stability. The design, with separate red and blue arms, is based on a dual R4 échelle grating with cross dispersion provided by Volume Phase Holographic Gratings. Separate fibres will permit simultaneous observation of target and sky; different fibre pairs and injection optics will provide three different resolving powers between $\sim$17 000 and 80 000. These properties are pertinent to a range of astrophysical problems of interest to both SALT partners and University of Canterbury astronomers.

INTRODUCTION

Astronomers at the University of Canterbury have a history of building and using high-resolution spectrographs at their Mount John University Observatory. The Cassegrain échelle spectrograph built by Hearnshaw in the 1970s (a copy of the Harvard design) was one of the first échelle spectrographs in the southern hemisphere [1]. In 2002 it was superseded by the HERCULES échelle spectrograph [2], notable features of which include a fibre feed and vacuum tank for wavelength stability and a cross-dispersing prism to preserve throughput.

In about 2000, wishing to capitalize on this experience, the SALT Science Working Group (SSWG) and the SALT Board made an in-principle decision to accept an offer from the University of Canterbury to build a high-resolution spectrograph (HRS) for SALT, for which Canterbury would make a US$500 000[1] non-cash, in-kind contribution.

TABLE 1. Science drivers for high-resolution spectroscopy with SALT.

- Element abundance studies and distance determinations in Local Group galaxies
- Distances and evolutionary states of hot eclipsing binaries in the Magellanic Clouds
- Stellar internal structure and dynamics
- Atmospheric dynamics of Cepheid variables
- Star cluster and Galaxy dynamics
- Outflow and accretion studies
- Extrasolar planet detection from radial-velocity variations
- Line-profile analyses in β Pictoris-like objects
- Target of opportunity follow-up of microlensing events
- High- and moderate-redshift galaxy spectroscopy

[1] Now increased to US$550 000.

CP752, *Stellar Astrophysics with the World's Largest Telescopes: First International Workshop*
edited by J. Mikołajewska and A. Olech

TABLE 2. Design parameters for the SALT high-resolution spectrograph.

- Simultaneous recording of a single object + sky
- Accurate sky subtraction (one sky fibre sufficient)
- Wavelength coverage 370–890 nm
- Resolving powers 17 000–80 000
- Radial-velocity precision $\sim$20 m s^{-1}

Some science drivers for such an instrument are listed in Table 1. Many of these align with Canterbury astronomers' interests, but SALT HRS will provide additional opportunities in other research fields.

SPECTROGRAPH DESIGN

Table 2 indicates design parameters agreed with the SSWG. The radial-velocity stability, essential for exoplanet detection, implies a HERCULES-like stationary instrument operating in a vacuum tank and fed by a light-scrambling fibre.

HERCULES employs an R2 échelle grating (i.e. one for which tan[blaze angle] = 2) and an initial design for a SALT HRS was produced based on this geometry [3]. However during a first Preliminary Design Review in September 2003 it was considered that this design was too risky and that the use of R4 gratings should be explored. The risks were due to the massive cross-dispersing prisms (two at $\sim$100 kg each) and large camera lenses (up to 800 mm in diameter). An R4 design did indeed produce smaller optical elements [4]. As a result of an excellent second Preliminary Design Review in July 2004, the SALT Project Scientist, Dr David Buckley, has recommended to the SALT Board that the R4 design should proceed to Critical Design Review in 2005. Although elements of the R4 design require further refinement, its essentials are now clear.

A Fibre Instrument Feed above SALT's Spherical Aberration Collector will present

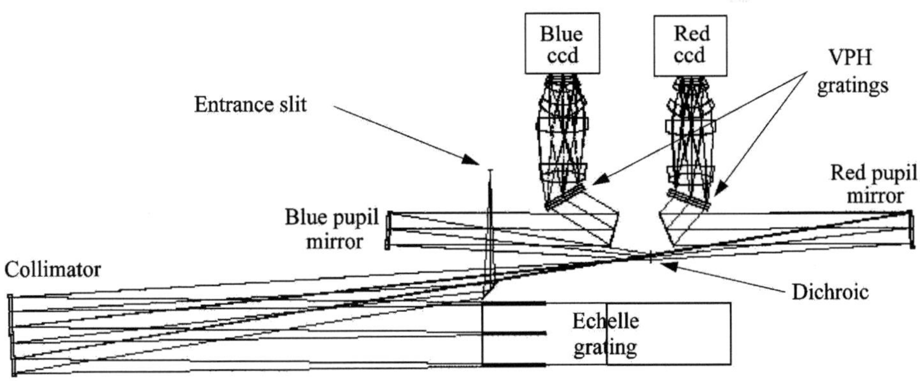

FIGURE 1. Post-PDR R4 spectrograph design. Key elements are annotated and described briefly in the text.

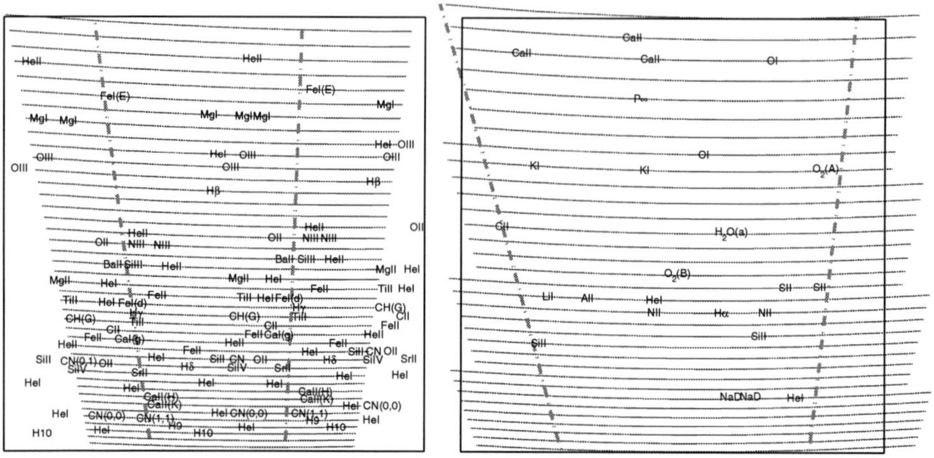

FIGURE 2. Layout of the blue (*left*) and red (*right*) CCD focal planes. The outer boxes indicate 60×60-mm squares characteristic of the detector size. The thick dot-dashed lines delimit the free spectral range. Important absorption features are marked.

target and sky fibres in the primary focal plane and can accommodate up to six fibre pairs. The separation of each fibre pair will be variable between 15 arcsec and 3.7 acrmin. Illumination for wavelength and smooth-field calibrations will be provided through the prime-focus calibration system common to all instruments. The fibre run to the spectrograph room directly beneath the telescope will be some 30 m. Immediately outside the spectrograph's vacuum enclosure injection optics will be necessary to convert the ~f/3.8 fibre outputs to the ~f/20 beams required by the Bowen-Walraven type image slicers needed for the two higher resolving power modes and then the ~f/10 beams for the collimator. Image slicers will not be needed for the lowest resolution mode.

The overall optical layout of the post-PDR R4 spectrograph is shown in Figure 1. The dispersing element will be two R4 échelle gratings replicated onto a single substrate. Dispersed light returns to the collimator mirror which then forms an intermediate focus after which a dichroic filter separates the beams into blue and red arms (370–550 nm and 555–890 nm, respectively). Each arm will comprise a convergent pupil mirror to form parallel beams that are fed to a cross-dispersing Volume Phase Holographic Grating (VPHG), camera optics and finally a cryogenic CCD detector. VPHGs work in transmission and diffract most of the light into a single order, preserving throughput. Figure 2 shows the layout of the CCD focal planes.

KiwiStar Optics Ltd of Lower Hutt near Wellington in New Zealand are the preferred optical fabricators, while mechanical construction and assembly will occur in the Department of Physics & Astronomy in Christchurch. Gratings, beam splitters and CCD systems will be sourced from external suppliers. Damien Jones, of Prime Optics in Queensland, Australia, is providing additional optical design expertise through a consultancy. Graham Hodge, of Unilogix Ltd in Christchurch, is providing project management through a similar mechanism.

Calculations show that across most of the spectral range the overall efficiency of

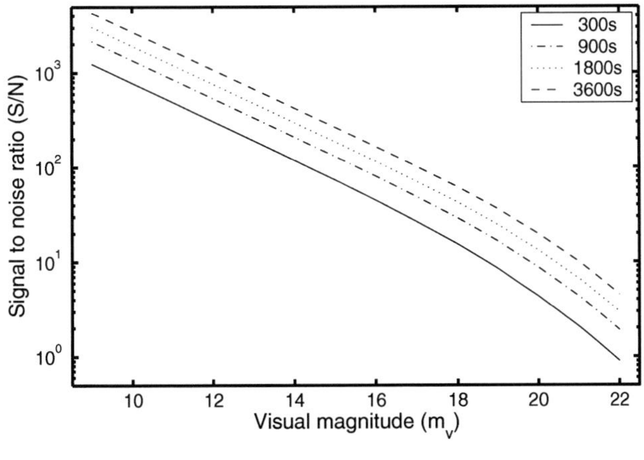

FIGURE 3. Expected signal-to-noise ratios for G0 V stars at an airmass of 1.3 and a wavelength of 650 nm for the $R \sim 17\,000$ mode. Exposure times range from 300 to 3 600 s.

the HRS (telescope + slit + spectrograph + CCD) will lie in the range of 10–15 per cent. Figure 3 translates these figures into a sample of expected performance. Possible refinements to the design will not affect these estimates significantly.

CONCLUSION

Thanks to a committed design team and productive comments at Preliminary Design Reviews, we are converging on an excellent HRS design. We expect that SALT Board go-ahead will lead to Critical Design Review in 2005 followed by fabrication, assembly and testing to early 2007, with commissioning of a state-of-the-art HRS on SALT later in that year.

REFERENCES

1. J.B. Hearnshaw, *Proc. Astron. Soc. Australia* **3,** 102-103 (1977).
2. J.B. Hearnshaw, S.I. Barnes, G.M. Kershaw, N. Frost, G. Graham, R. Ritchie and G.R. Nankivell, *Experimental Astron.* **13,** 59-76 (2002).
3. S.I. Barnes, J.B. Hearnshaw, P.L. Cottrell, M.D. Albrow and G. Kershaw, *Proceedings of SPIE-Volume 4841: Instrument Design and Performance for Optical/Infrared Ground-based Telescopes,* 1157-1161 (2003).
4. S.I. Barnes, P.L. Cottrell, M.D. Albrow and G. Kershaw, *SALT HRS 3210AE0001 Optical Design Definition Document* (2003).

STAR FORMATION AND PLANETARY SYSTEMS

The Transiting Planets

Tsevi Mazeh

Tel Aviv University, Tel Aviv Israel

Abstract.
The paper points out the importance of the recently discovered transiting planets and discusses the photometric searches for such extrasolar giant planets. The seven known transiting planets suggest a *possible* correlation between the masses and periods of the close-in giant planets.

INTRODUCTION

One of the most important breakthroughs in Astronomy of the last few decades was the discovery of extra-solar giant planets (Mayor & Queloz 1995). Planets were discovered by detecting the minute motion of their parent stars induced by the gravitational attraction of the planets (e.g., Marcy & Butler 1996), a motion that can be detected by careful monitoring of the stellar radial velocity. The discoveries of the last decade became possible when the stellar speedometers reached a precision of the order of 10 m/s. Note, however, that an early detection of a planet *candidate* around HD 114762, with a mass of about 10 M_{Jup}, was reported already by Latham et al (1989), with a precision of 500 m/s.

The recent discoveries revealed new planetary features, totally unexpected by our previous naive notion of planetary systems, which was based solely on the Solar system. The main feature is the existence of short-period giant planets in close orbits around their host stars. As shown in Figure 1, more than 20 planets, which consist of more than 15% of all detected giant planets discovered by radial velocity measurements, have periods shorter than 10 days and therefore orbital radii smaller than 0.1 AU. This feature is very different from what we know about our solar system, where the giant planets all have orbital radii larger than 5 AU. The generally accepted paradigm before the discovery of the extra-solar planets assumed that the giant planets could not have been formed in small distances from their parent stars. Therefore, theoretical models to explain the formation, evolution and survival of the close-in giant planets are still under hot debate (e.g., Sari & Goldreich 2004; Laughlin, Steinacker & Adams 2004; Boss 2004).

As of August 2004, more than 120 planet candidates have been found around nearby solar-type stars (e.g., Schneider 2004; Marcy et al. 2004). Although they provided valuable new information about the orbital parameters of the population of extrasolar planets (Marcy et al. 2003; Udry et al. 2003) and the host star properties (Santos et al. 2003; Laws et al. 2003), the observations did not yield the true masses for most of the planets, because the orbital inclinations can not be derived. So far, the actual masses could be derived for one planet around nearby star, Gls 876, with astrometric orbit detected by HST measurements (Benedict et al. 2002), and for planets that exhibit transits, indicating orbital inclinations close to 90°. The detailed light curve of the transit

CP752, *Stellar Astrophysics with the World's Largest Telescopes: First International Workshop*
edited by J. Mikołajewska and A. Olech

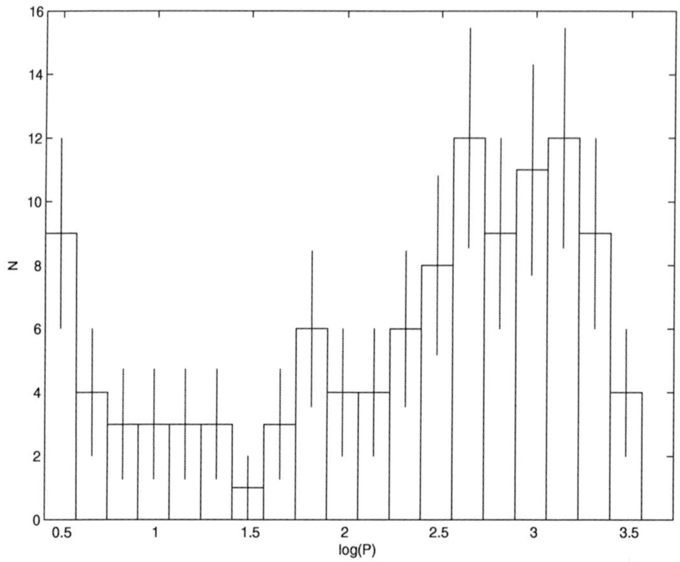

FIGURE 1. Period histogram of the extrasolar planets detected by RV

can yield an accurate estimate of the inclination. Together with the radial-velocity curves and some reasonable assumptions about the stellar masses, the inclination angles enable us to derive the planetary masses. The planetary radii can also be obtained only for the transiting planets, so we can derive their mean density and surface gravity.

HD 209458

The first transiting planet to be discovered was HD209458B (Charbonneau et al. 2000; Henry et al. 2000), after the radial-velocity modulation of its parent star (Mazeh et al. 2000) had indicated a planet with minimum mass of about 0.7 M_{Jup}.

Further observations of HD209458 with HST (Brown et al. 2001) produced an unprecedentedly precise light curve of the transit, which enabled Brown et al. to derive the planetary radius and orbital inclination without having to assume a radius for the parent star. The precise photometry has been even used to put an upper limit to the size of possible moons and rings around HD209458b!

The discovery of the planetary transit initiated novel studies of the system. Queloz et al. (2000) have detected a 'Rossiter effect' — a distortion of the stellar line profiles of HD209458 during the planetary transit, caused by the planet crossing the disk of the rotating star. They found that the planetary orbital motion is in the same direction as the stellar rotation, and even succeeded to estimate the inclination angle of the orbital plane relative to the stellar equatorial plane, which they find as small as 3°.

Finally, transiting extrasolar planets present a unique opportunity to learn about the atmosphere of these objects. Spectroscopic observations of HD209458 with HST detected the sodium resonant doublet in the planetary atmosphere (Charbonneau et al. 2002), escaping atomic hydrogen (Vidal-Madjar et al. 2003) and C I and O II in the planetary extended atmosphere (Vidal-Madjar et al. 2004). This is a fantastic achievement, given the fact that the parent star shines in the immediate vicinity of the planet. The study of the planetary atmosphere probably does not have to be from space, as suggested by Snellen (2004).

PHOTOMETRIC SEARCH FOR TRANSITING PLANETS

The great success of the observations of HD209458 transit drove quite a few teams to search for additional transiting planets by the opposite approach. In this approach a large number of stars are monitored photometrically in search of transiting light curve (e.g., Deeg et al. 1998; Gilliland et al. 2000). For parent stars of G or K main-sequence stars, such light curves are characterized by flat, $\sim$1% depth, eclipse. The existence of the planets is confirmed by follow-up radial-velocity measurements.

The new ground-based projects (see Horne 2003 for a review) can be divided schematically into two broad categories:

- Projects that monitor small but crowded fields on the sky, like the Galactic bulge or a globular cluster, typically with a 1–4 m telescope (e.g., Udalski et al. 2002a,b; Burke et al. 2004; Yee et al. 2003; Mochejska et al. 2004). Such projects monitor a few tens of thousands of stars, all concentrated in a field of view (=FoV) of less than 1 squared degree. The typical limiting magnitude of these projects is $\sim$ 16 mag.
- Projects that monitor only relatively bright stars with small, dedicated telescopes with apertures of a few inches (e.g., Hartman et al. 2004; Borucki et al. 2001; Kane et al. 2004; Alonso et al. 2004). Such projects observe large FoVs, tens of squared degrees each, with a typical limiting magnitude of 12–14 mag.

In addition, two space missions have already been approved for transit search: CoRoT (e.g., Bordé, Rouan & Léger 2003) and Kepler (e.g., Borucki et al. 2004). Both missions will be launched within a few years.

A typical transit-candidate star detected by the 1-m class telescopes is rather faint. For example, the stars found by OGLE 1.3-m telescope (Udalski et al. 2002a,b; 2003) are in the range of 14–16 mag, rendering follow-up observations difficult. On the other hand, the small telescopes can detect transit-like light curves only for stars brighter than 12-13 mag, which are much easier for follow-up spectroscopic observations.

As of October 2004, only two photometric surveys yielded confirmed transiting planets. The reason for the negative results might be (see Brown 2003):

- Low detection efficiency of the presently running projects, because the precision of the individual measurements is lower than expected and the number of observations is too small.
- Low frequency of late-type stars included in the photometric samples. The available precision can not detect transits around stars of F spectral type or earlier.

- Low frequency of planets around stars that are members of wide stellar binaries, which are inevitably included in the photometric surveys. Such systems were avoided by the radial-velocity surveys, thus biasing upward the estimated frequency of planets.
- Low frequency of planets around solar-metallicity stars (Laws et al. 2003; Santos, Israelian & Mayor 2004) which is the most frequent metallicity of the nearby stars.

It is difficult to assess the relative contribution of each of the factors to the presently poor results of the photometric searches. Probably, the insufficient precision and the small number of measurements are major factors in most of the presently active wide-field photometric searches. In any case, the small number of new detections makes it even more desirable to detect photometrically more candidates for transiting planets. SALT with its high-resolution spectrograph will play a key role in confirming the planetary nature of these candidates.

Obviously, in order to detect shallow transits one needs to apply the best photometric reduction algorithm (e.g., Wozniak 2000), and to clean the photometric data from any systematic effect, as was done recently by Kruszewski & Semeniuk (2003) to the OGLE data. The next section presents a new algorithm that removes systematic effects in large photometric data.

SYS-REM: A NEW ALGORITHM TO REMOVE SYSTEMATIC EFFECTS

Very recently Tamuz, Mazeh & Zucker (2004) suggested a powerful algorithm — Sys-rem, that can remove from a large set of light curves any systematic effect, like the ones associated with atmospheric extinction, detector efficiency, or PSF changes over the detector. Sys-rem works without any prior knowledge of the effect, as long as it linearly appears in many stars of the sample. The algorithm, which was originally developed to remove atmospheric extinction effects, is based on a lower-rank approximation of matrices.

Originally, Sys-rem was developed to derive the best-fitting airmasses of the different images and the extinction coefficients of the different stars, without having any information on the stellar colours. However, the result is a general algorithm that deals with linear systematic effects. It turned out that such an algorithm had already been proposed and applied to data from disciplines other than astronomy, chemometrics, for example (Gabriel & Zamir 1979). In some restricted cases, when one can ignore the different uncertainties of the data points, this algorithm reduces to the well-known Principal Component Analysis. However, when the uncertainties of the measurements vary substantially, like in many photometric surveys, PCA performs poorly.

Sys-rem was applied to the Carina data of OGLE (Udalski et al. 2002b) with encouraging preliminary results. Figure 2 is the averaged periodogram of 200 randomly chosen stars from the Carina data. The upper panel shows the results of the original data, while the lower panel presents the result after Sys-rem was applied. The original periodogram clearly shows peaks at one-day frequency and its harmonics. Those peaks strongly indicate that the data still include systematic effects with a period of one day, which could

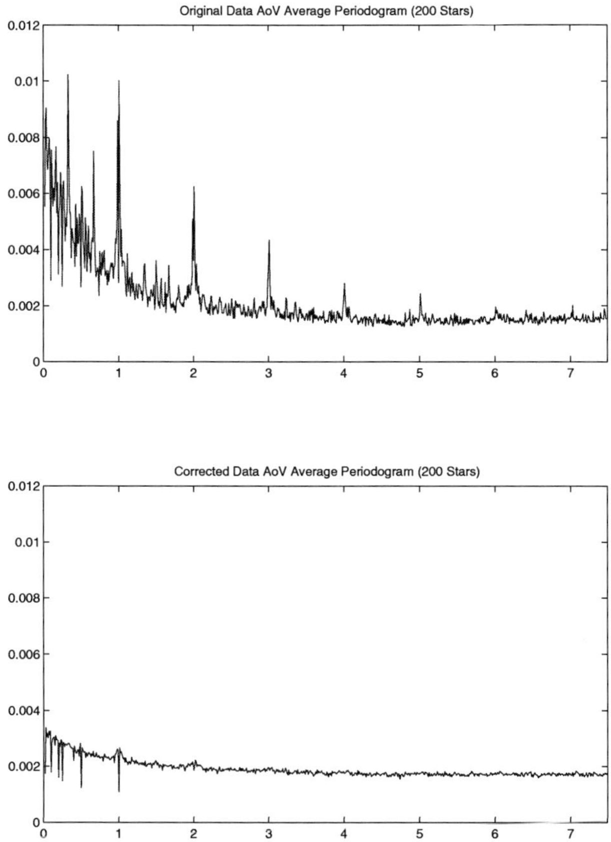

FIGURE 2. The averaged periodogram of 200 stars from the OGLE Carina field, before and after Sys-rem was applied

come from extinction absorption, for example. Sys-rem got rid of all these effects. In addition, the low-frequency rise presented in the original data almost disappeared after Sys-rem was applied.

THE KNOWN TRANSITING PLANETS

The presently most successful search for planets is the OGLE high-quality photometric survey (Udalski et al. 2002a,b; 2003), which yielded, after applying the BLS transit-search algorithm (Kovacs, Zucker & Mazeh 2002), more than one hundred transit candidates. Follow-up radial-velocity observations confirmed that OGLE-TR-56 (Konacki et al. 2003; Torres et al. 2004), OGLE-TR-113 (Bouchy et al. 2004; Konacki et al. 2004a) OGLE-TR-132 (Bouchy et al. 2004) and OGLE-TR-111 (Pont et al. 2004) all have planetary companions. Very recently, one detection by a wide-field small telescope, TrES-

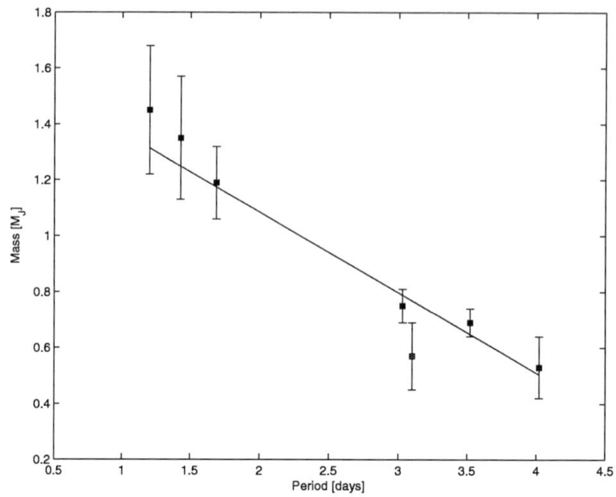

FIGURE 3. The masses and periods of the seven known transiting planets, with their best-fit linear dependence.

1, was announced (Alonso et al. 2004). Finally, Konacki et al. (2004b) confirmed the planetary nature of OGLE-TR-10.

The orbital periods of the seven transiting planets are 1.2, 1.43, 1.69, 3.03, 3.10, 3.52 and 4.02 days. The very short periods of the first three planets are challenging the theoreticians. They try to explain the evolution and survival of such close-in giant planets, which are under the intense radiation of their parent stars (e.g., Lammer et al. 2003; Burrows et al. 2004; Chabrier et al. 2004; Baraffe et al. 2004; Lecavelier des Etangs et al. 2004).

Mazeh, Zucker & Pont (2004) pointed out very recently to an intriguing correlation between the masses of the transiting planets and their periods, based on the first six transiting planets. Since that paper was published one more transit candidate — OGLE-TR-10, was confirmed. Figure 3 presents the seven known planetary masses as a function of their periods, and their best linear fit. Despite the fact that we have only seven points, the possible linearity of their positions in the diagram is intriguing. Mazeh, Zucker & Pont (2004) showed that the minimum masses known for the other short period planets are consistent with the possible correlation.

The suggested mass-period relation is based only on the planets with periods shorter than 5 days. For longer periods, the planetary masses seem to be correlated with the periods in the opposite direction (Zucker & Mazeh 2002). Therefore, the suggested correlation is based on very small number statistics, and more points are certainly needed to establish its existence. Nevertheless, the next section speculates on the possible mechanism behind such an intriguing feature.

Lecavelier des Etangs et al. (2004, see also Lammer et al. 2003 and Chabrier et al. 2004) calculated the evaporation rates of the atmosphere of close-in giant planets, taking into account thermal evaporation caused by the fact that the exobase of the atmosphere reaches the Roche limit of the planet. Consequently, planets with masses below some critical mass can not survive for too long. Such a process can cleanse the area in the mass-period diagram below the line that represents the critical mass. Clearly, the critical mass gets smaller as the orbital distance gets larger. Therefore, such a mechanism can explain the feature seen in the diagram. Moreover, planets positioned on the suggested mass-period line might have suffered an increase of their radius because of the stellar heating, and would therefore be the easiest to detect by transit search. This might account for the fact that all the transiting planets are so close to the line in the diagram.

Mazeh, Zucker and Pont (2004) pointed out a few open questions of such a model. In particular, one would think that the evaporation process should depend strongly on the stellar brightness, a feature which is not seen in the actual data of the short-period planets. Therefore we need many more transiting planets to see if the correlation still holds, and how the planetary mass depends on the other parameters of the system. The intense photometric surveys will certainly produce many such transits in the near future.

REFERENCES

1. R. Alonso, T. M. Brown, G. Torres, D. W. Latham, A. Sozzetti, G. Mandushev, J. A. Belmonte, D. Charbonneau, H. J. Deeg, E. W. Dunham, F. T. O'Donovan, and R. P. Stefanik. TrES-1: The Transiting Planet of a Bright K0 V Star. *Ap.J. Lett.*, 613:L153–L156, October 2004.
2. I. Baraffe, F. Selsis, G. Chabrier, T. S. Barman, F. Allard, P. H. Hauschildt, and H. Lammer. The effect of evaporation on the evolution of close-in giant planets. *A&A*, 419:L13–L16, May 2004.
3. G. F. Benedict, B. E. McArthur, T. Forveille, X. Delfosse, E. Nelan, R. P. Butler, W. Spiesman, G. Marcy, B. Goldman, C. Perrier, W. H. Jefferys, and M. Mayor. A Mass for the Extrasolar Planet Gliese 876b Determined from Hubble Space Telescope Fine Guidance Sensor 3 Astrometry and High-Precision Radial Velocities. *Ap.J. Lett.*, 581:L115–L118, December 2002.
4. P. Bordé, D. Rouan, and A. Léger. Exoplanet detection capability of the COROT space mission. *A&A*, 405:1137–1144, July 2003.
5. W. Borucki, D. Koch, A. Boss, E. Dunham, A. Dupree, J. Geary, R. Gilliland, S. Howell, J. Jenkins, Y. Kondo, D. Latham, J. Lissauer, and H. Reitsema. . In *Second Eddington Workshop: Stellar structure and habitable planet finding, 9 - 11 April 2003, Palermo, Italy. Edited by F. Favata, S. Aigrain and A. Wilson. ESA SP-538, Noordwijk: ESA Publications Division, ISBN 92-9092-848-4*, page 177, January 2004.
6. W. J. Borucki, D. Caldwell, D. G. Koch, L. D. Webster, J. M. Jenkins, Z. Ninkov, and R. Showen. The Vulcan Photometer: A Dedicated Photometer for Extrasolar Planet Searches. *PASP*, 113:439–451, April 2001.
7. A. P. Boss. Convective Cooling of Protoplanetary Disks and Rapid Giant Planet Formation. *ApJ.*, 610:456–463, July 2004.
8. F. Bouchy, F. Pont, N. C. Santos, C. Melo, M. Mayor, D. Queloz, and S. Udry. Two new "very hot Jupiters" among the OGLE transiting candidates. *A&A*, 421:L13–L16, July 2004.
9. T. M. Brown. Expected Detection and False Alarm Rates for Transiting Jovian Planets. *Ap.J. Lett.*, 593:L125–L128, August 2003.
10. T. M. Brown, D. Charbonneau, R. L. Gilliland, R. W. Noyes, and A. Burrows. Hubble Space Telescope Time-Series Photometry of the Transiting Planet of HD 209458. *ApJ.*, 552:699–709, May 2001.
11. C. J. Burke, B. S. Gaudi, D. L. DePoy, R. W. Pogge, and M. H. Pinsonneault. Survey for Transiting Extrasolar Planets in Stellar Systems. I. Fundamental Parameters of the Open Cluster NGC 1245. *Astron. J.*, 127:2382–2397, April 2004.
12. A. Burrows, I. Hubeny, W. B. Hubbard, D. Sudarsky, and J. J. Fortney. Theoretical Radii of Transiting Giant Planets: The Case of OGLE-TR-56b. *Ap.J. Lett.*, 610:L53–L56, July 2004.
13. G. Chabrier, T. Barman, I. Baraffe, F. Allard, and P. H. Hauschildt. The Evolution of Irradiated Planets: Application to Transits. *Ap.J. Lett.*, 603:L53–L56, March 2004.

14. D. Charbonneau, T. M. Brown, D. W. Latham, and M. Mayor. Detection of Planetary Transits Across a Sun-like Star. *Ap.J. Lett.*, 529:L45–L48, January 2000.
15. D. Charbonneau, T. M. Brown, R. W. Noyes, and R. L. Gilliland. Detection of an Extrasolar Planet Atmosphere. *ApJ.*, 568:377–384, March 2002.
16. H. J. Deeg, L. R. Doyle, V. P. Kozhevnikov, E. L. Martin, B. Oetiker, E. Palaiologou, J. Schneider, C. Afonso, E. W. Dunham, J. M. Jenkins, Z. Ninkov, R. P. S. Stone, and P. E. Zakharova. Near-term detectability of terrestrial extrasolar planets: TEP network observations of CM Draconis. *A&A*, 338:479–490, October 1998.
17. K. R. Gabriel and S. Zamir. Lower Rank Approximation of Matrices by Least Squares with Any Choice of Weights. *Technometrics*, 21:489–498, November 1979.
18. R. L. Gilliland, T. M. Brown, P. Guhathakurta, A. Sarajedini, E. F. Milone, M. D. Albrow, N. R. Baliber, H. Bruntt, A. Burrows, D. Charbonneau, P. Choi, W. D. Cochran, P. D. Edmonds, S. Frandsen, , and and 8 coauthors,. A Lack of Planets in 47 Tucanae from a Hubble Space Telescope Search. *Ap.J. Lett.*, 545:L47–L51, December 2000.
19. J. D. Hartman, G. Bakos, K. Z. Stanek, and R. W. Noyes. HATNET Variability Survey in the High Stellar Density "Kepler Field" with Millimagnitude Image Subtraction Photometry. *Astron. J.*, 128:1761–1783, October 2004.
20. G. W. Henry, G. W. Marcy, R. P. Butler, and S. S. Vogt. A Transiting "51 Peg-like" Planet. *Ap.J. Lett.*, 529:L41–L44, January 2000.
21. K. Horne. . In *ASP Conf. Ser. 294: Scientific Frontiers in Research on Extrasolar Planets*, page 361, 2003.
22. S. R. Kane, A. Collier Cameron, K. Horne, D. James, T. A. Lister, D. L. Pollacco, R. A. Street, and Y. Tsapras. Results from the Wide-Angle Search for Planets prototype (WASP0) - I. Analysis of the Pegasus field. *MNRAS*, 353:689–696, September 2004.
23. M. Konacki and et al. astro-ph/0412400, 2004.
24. M. Konacki, G. Torres, S. Jha, and D. D. Sasselov. An extrasolar planet that transits the disk of its parent star. *Nature*, 421:507–509, January 2003.
25. M. Konacki, G. Torres, D. D. Sasselov, G. Pietrzyński, A. Udalski, S. Jha, M. T. Ruiz, W. Gieren, and D. Minniti. The Transiting Extrasolar Giant Planet around the Star OGLE-TR-113. *Ap.J. Lett.*, 609:L37–L40, July 2004.
26. G. Kovács, S. Zucker, and T. Mazeh. A box-fitting algorithm in the search for periodic transits. *A&A*, 391:369–377, August 2002.
27. A. Kruszewski and I. Semeniuk. A Simple Method of Correcting Magnitudes for the Errors Introduced by Atmospheric Refraction. *Acta Astronomica*, 53:241–248, September 2003.
28. H. Lammer, F. Selsis, I. Ribas, E. F. Guinan, S. J. Bauer, and W. W. Weiss. Atmospheric Loss of Exoplanets Resulting from Stellar X-Ray and Extreme-Ultraviolet Heating. *Ap.J. Lett.*, 598:L121–L124, December 2003.
29. D. W. Latham, R. P. Stefanik, T. Mazeh, M. Mayor, and G. Burki. The unseen companion of HD114762 - A probable brown dwarf. *Nature*, 339:38–40, May 1989.
30. G. Laughlin, A. Steinacker, and F. C. Adams. Type I Planetary Migration with MHD Turbulence. *ApJ.*, 608:489–496, June 2004.
31. C. Laws, G. Gonzalez, K. M. Walker, and et al. Parent Stars of Extrasolar Planets. VII. New Abundance Analyses of 30 Systems. *Astron. J.*, 125:2664–2677, May 2003.
32. A. Lecavelier des Etangs, A. Vidal-Madjar, J. C. McConnell, and G. Hébrard. Atmospheric escape from hot Jupiters. *A&A*, 418:L1–L4, April 2004.
33. G. W. Marcy and R. P. Butler. A Planetary Companion to 70 Virginis. *Ap.J. Lett.*, 464:L147, June 1996.
34. G. W. Marcy, R. P. Butler, D. A. Fischer, and S. S. Vogt. . In *ASP Conf. Ser. 294: Scientific Frontiers in Research on Extrasolar Planets*, page 1, 2003.
35. G. W. Marcy, R. P. Butler, S. S. Vogt, and D. A. Fischer. . In *IAU Symposium*, page 11, June 2004.
36. G.W. Marcy and et al. http://exoplanets.org, 2004.
37. M. Mayor and D. Queloz. A Jupiter-Mass Companion to a Solar-Type Star. *Nature*, 378:355, November 1995.
38. T. Mazeh, D. Naef, G. Torres, D. W. Latham, M. Mayor, J. Beuzit, T. M. Brown, L. Buchhave, M. Burnet, B. W. Carney, D. Charbonneau, G. A. Drukier, J. B. Laird, F. Pepe, C. Perrier, D. Queloz, N. C. Santos, J. Sivan, S. Udry, and S. Zucker. The Spectroscopic Orbit of the Planetary Companion

Transiting HD 209458. *Ap.J. Lett.*, 532:L55–L58, March 2000.
39. T. Mazeh, S. Zucker, and F. Pont. An intriguing correlation between the masses and periods of the transiting planets. *MNRAS*, pages 671–+, November 2004.
40. B. J. Mochejska, K. Z. Stanek, D. D. Sasselov, A. H. Szentgyorgyi, M. Westover, and J. N. Winn. Planets in Stellar Clusters Extensive Search. II. Discovery of 57 Variables in the Cluster NGC 2158 with Millimagnitude Image Subtraction Photometry. *Astron. J.*, 128:312–322, July 2004.
41. F. Pont, F. Bouchy, D. Queloz, N. C. Santos, C. Melo, M. Mayor, and S. Udry. The "missing link": A 4-day period transiting exoplanet around OGLE-TR-111. *A&A*, 426:L15–L18, October 2004.
42. D. Queloz, A. Eggenberger, M. Mayor, C. Perrier, J. L. Beuzit, D. Naef, J. P. Sivan, and S. Udry. Detection of a spectroscopic transit by the planet orbiting the star HD209458. *A&A*, 359:L13–L17, July 2000.
43. N. C. Santos, G. Israelian, and M. Mayor. Spectroscopic [Fe/H] for 98 extra-solar planet-host stars. Exploring the probability of planet formation. *A&A*, 415:1153–1166, March 2004.
44. R. Sari and P. Goldreich. Planet-Disk Symbiosis. *Ap.J. Lett.*, 606:L77–L80, May 2004.
45. J. Schneider. http://www.obspm.fr/encycl/encycl.html, 2004.
46. I. A. G. Snellen. A new method for probing the atmospheres of transiting exoplanets. *MNRAS*, 353:L1–L6, September 2004.
47. O. Tamuz, T. Mazeh, and S. Zucker. Correcting systematic effects in a large set of photometric light curves. *MNRAS*, pages 712–+, December 2004.
48. G. Torres, M. Konacki, D. D. Sasselov, and S. Jha. New Data and Improved Parameters for the Extrasolar Transiting Planet OGLE-TR-56b. *ApJ.*, 609:1071–1075, July 2004.
49. A. Udalski, B. Paczynski, K. Zebrun, M. Szymaski, M. Kubiak, I. Soszynski, O. Szewczyk, L. Wyrzykowski, and G. Pietrzynski. The Optical Gravitational Lensing Experiment. Search for Planetary and Low-Luminosity Object Transits in the Galactic Disk. Results of 2001 Campaign. *Acta Astronomica*, 52:1–37, March 2002.
50. A. Udalski, G. Pietrzynski, M. Szymanski, M. Kubiak, K. Zebrun, I. Soszynski, O. Szewczyk, and L. Wyrzykowski. The Optical Gravitational Lensing Experiment. Additional Planetary and Low-Luminosity Object Transits from the OGLE 2001 and 2002 Observational Campaigns. *Acta Astronomica*, 53:133–149, June 2003.
51. A. Udalski, O. Szewczyk, K. Zebrun, G. Pietrzynski, M. Szymanski, M. Kubiak, I. Soszynski, and L. Wyrzykowski. The Optical Gravitational Lensing Experiment. Planetary and Low-Luminosity Object Transits in the Carina Fields of the Galactic Disk. *Acta Astronomica*, 52:317–359, December 2002.
52. S. Udry, M. Mayor, and N. C. Santos. Statistical properties of exoplanets. I. The period distribution: Constraints for the migration scenario. *A&A*, 407:369–376, August 2003.
53. A. Vidal-Madjar, J.-M. Désert, A. Lecavelier des Etangs, G. Hébrard, G. E. Ballester, D. Ehrenreich, R. Ferlet, J. C. McConnell, M. Mayor, and C. D. Parkinson. Detection of Oxygen and Carbon in the Hydrodynamically Escaping Atmosphere of the Extrasolar Planet HD 209458b. *Ap.J. Lett.*, 604:L69–L72, March 2004.
54. A. Vidal-Madjar, A. Lecavelier des Etangs, J.-M. Désert, G. E. Ballester, R. Ferlet, G. Hébrard, and M. Mayor. An extended upper atmosphere around the extrasolar planet HD209458b. *Nature*, 422:143–146, March 2003.
55. P. R. Wozniak. Difference Image Analysis of the OGLE-II Bulge Data. I. The Method. *Acta Astronomica*, 50:421–450, December 2000.
56. H. K. C. Yee, G. Mallen-Ornelas, S. Seager, M. Gladders, T. X. Brown, D. Minnitti, S. Ellison, and G. Mallen-Fullerton. . In *Discoveries and Research Prospects from 6- to 10-Meter- Class Telescopes II. Edited by Guhathakurta, Puragra. Proceedings of the SPIE, Volume 4834*, page 150, February 2003.
57. S. Zucker and T. Mazeh. On the Mass-Period Correlation of the Extrasolar Planets. *Ap.J. Lett.*, 568:L113–L116, April 2002.

Radial velocity measurements of a sample of K-giants with the Hobby-Eberly telescope.

Niedzielski, A.*, Wolszczan, A.† and Konacki, M.**

*Torun Centre for Astronomy, Nicolaus Copernicus University, Gagarina 11, 87-100 Torun, Poland; aniedzi@astri.uni.torun.pl
†Pennsylvania State University, Department of Astronomy and Astrophysics, 525 Davey Laboratory, University Park, PA 16802, USA; alex@astro.psu.edu
**California Institute of Technology, Div. of Geological and Planetary Sciences, MC 150-21, Pasadena, CA 91125, USA; maciej@gps.caltech.edu

Abstract. We present motivation and initial results of a large RV survey of K giants aimed at a detection of low-mass companions. The survey, performed with the Hobby-Eberly Telescope, utilizes high resolution (60,000) spectra for high precision radial velocity measurements. The primary goal of the survey is the selection of astrometrically stable reference stars for the Extra-solar Planet Interferometric Survey key project to be carried out with the Space Interferometry Mission.

INTRODUCTION

The main goal of this program is to select astrometric reference stars for a subset of 24 target stars to be searched for Earth-mass companions by the Space Interferometry Mission (SIM) key project *Extra-solar Planet Interferometric Survey* (EPIcS, [12]). (*http://planetquest.jpl.nasa.gov/SIM*). A $\sim$1 μas precision of the survey calls for reference stars that will keep the astrometric noise below that level. They also have to be prevalent enough to be easily found within 1° of a target for narrow-angle astrometry. These conditions appear to be best satisfied by clump giants [8] at distances $\sim$1 kpc away, because of the $R < 10^m$ SIM requirement. As this distance is insufficient to completely eliminate the possibility of an astrometric contamination by giant planets, each target requires two reference stars. About 10 reference star candidates per target (early K-giants) have been identified. These stars have to be radial velocity (RV) monitored to select the best two on the basis of the absence of a detectable binary motion signature.

Aside from the SIM/EPIcS reference star selection to be accomplished by this program, the precision RV measurements of a set of $\sim$300 distant K-giants accumulated in the selection process will provide a statistically significant and astrophysically very interesting material for more detailed investigations.

One of the most intriguing properties of red giants is their RV variation. [14] noted that all giants in their (small) sample display RV variations of 30-300 m/s. Long-term (>100 day) variations with RV amplitudes of several hundred m/s, reported by [5], may be induced by stellar rotation influencing active zones on stellar surface or by the existence of low-mass companions. Yet another type of RV variations, reported by [6],[7], the

CP752, *Stellar Astrophysics with the World's Largest Telescopes: First International Workshop*
edited by J. Mikołajewska and A. Olech

short-term RV changes of about 100 m/s within a ~day are probably related to high order radial- or non-radial pulsations. [3] has reported the first results of a high precision RV survey of 182 K giants and presented evidence of a <50 m/s RV scatter in these stars that is most probably of an intrinsic origin.

It is interesting to note that, in spite of these uncertainties in RVs of K giants, several sub-stellar companions have already been found around these stars. First such companion was reported to orbit ιDra by [4]. More such detections have been reported by [13] (HD 47536), [11] (HD 59686, 91 Aqr, τ Gem, ν Oph). Finally, two sub-stellar companions to sub-giants have been reported by [1], [2].

The ongoing precision RV surveys usually involve main sequence stars (dwarfs) [10]. However, since the existence of planets orbiting evolved stars (giants) is extremely interesting from the point of view of the evolution of planetary systems, we have expanded the original goals of the SIM/EPIcS survey to include a detection of planets around the K-giant candidates for EPIcS reference stars.

OBSERVATIONS

Observations have been performed with the Hobby-Eberly Telescope (HET) located at the McDonald Observatory, Texas, using the High Resolution Spectrograph (HRS) at 60,000 resolution. The observing scheme follows standard practices implemented in high precision radial velocity measurements with the iodine (I2) cell [9].

Our program has been initiated in January 2004. In two trimesters of HET observations, between January and August 2004, we have collected 271 spectra for 120 objects. Observations for several stars were repeated at a several months interval to test the data reduction procedure and the performance of the RV measurement code. Altogether, we collected repeated observations for 26 stars. Ten of them were repeated in 2-6 months for tests, while the remaining 16 were repeated to improve the signal-to-noise of the spectra.

INITIAL RESULTS OF RADIAL VELOCITY MEASUREMENTS

Multiple observations of several K-giant EPIcS reference candidates were reduced with the standard IRAF [1] procedures. Radial velocities were calculated using the standard I2 RV technique [9]. The initial results are presented in Figure 1. A $\pm$50 m/s threshold has been tentatively adopted as a rejection criterion for candidate reference stars. The individual observations are plotted against epoch with error bars of $\pm 1\sigma$. The precision of our RV determinations varies in individual cases over a 19-32 m/s interval.

In Figure 1 (upper panel) an example of a binary system is presented. This bright giant of K0 spectral type was re-observed 12 and 143 days after the initial measurement. Our

[1] IRAF is distributed by the National Optical Astronomy Observatories, which are operated by the Association of Universities for Research in Astronomy, Inc., under cooperative agreement with the National Science Foundation

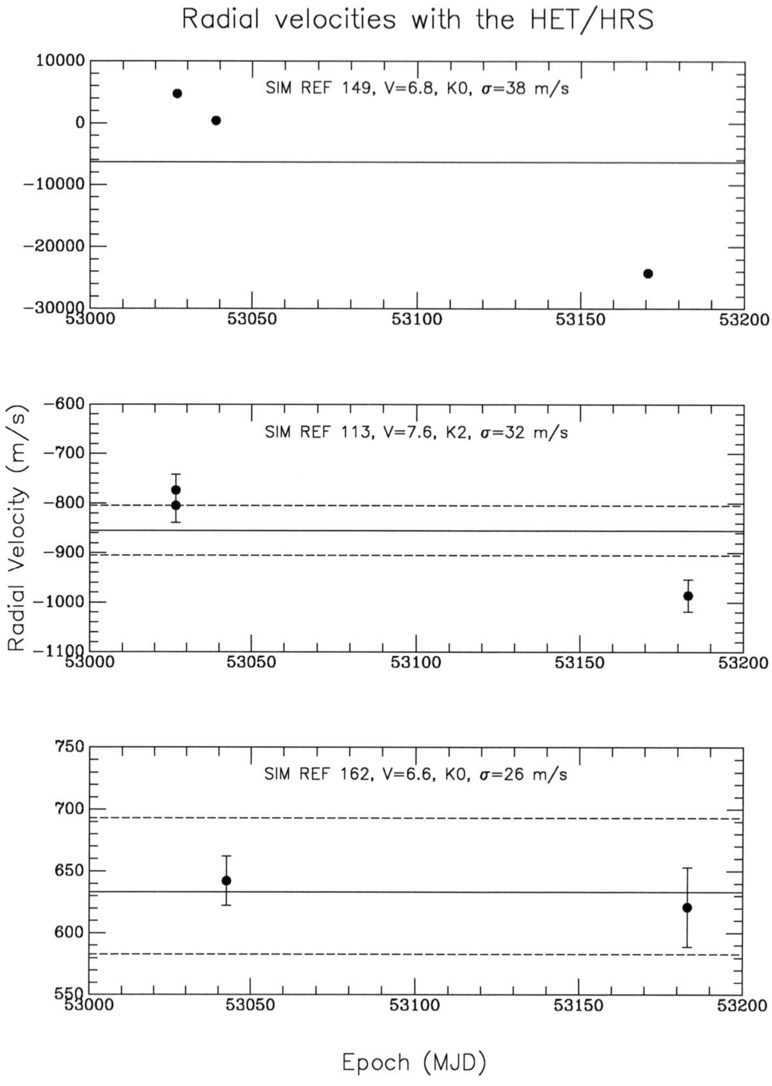

FIGURE 1. Relative radial velocities of three SIM/EPIcS candidate reference stars. The mean RV (relative to template) is plotted as solid line, dashed lines show the ±50 m/s threshold adopted for SIM/EPIcS reference stars pre-selection.

test calculations show that, in this case, RV variations in reach a 29 km/s level allowing us to reject the object from the SIM/EPIcS candidate list. In Figure 1 (middle panel) we show a system displaying much lower RV variations. Our observations of this bright K2 star repeated on the same night and after 157 days reveal RV variations at a 181 m/s level. This makes the star useless as a SIM/EPIcS reference object but it may be interesting as a possible host to a sub-stellar companion. Finally, in Figure 1 (bottom panel) we present an example of a promising SIM/EPIcS reference candidate. A bright star of K0 spectral type was re-observed after 141 days. Our calculations show that the RV changed by only 19 m/s, well within the uncertainty level accepted in the pre-selection process.

CONCLUSIONS

The queue-scheduled observing mode with the HET seems to be a very efficient way to implement a large, long-term project described above. The RV precision reached in our test measurements is already sufficient for SIM/EPIcS reference star pre-selection process. Our reduction procedure requires refinement to reach the RV precision adequate for finding Jovian-mass companions to our target stars.

ACKNOWLEDGMENTS

Kind cooperation of the Hobby-Eberly Telescope staff is acknowledged with thanks. We also thank C. Gelino for making available his list of SIM/EPIcS candidate reference stars. The Hobby-Eberly Telescope (HET) is a joint project of the University of Texas at Austin, the Pennsylvania State University, Stanford University, Ludwig-Maximilians-Universität München, and Georg-August-Universität Göttingen. The HET is named in honor of its principal benefactors, William P. Hobby and Robert E. Eberly.

REFERENCES

1. Butler, R. P., Tinney, C. G., Marcy, G. W., Jones, H. R. A., Penny, A. J., Apps, K. 2001 ApJ 555, 410
2. Cochran, W.D., Hatzes, A. P., Endl, M., Paulson, D.B., Walker, G. A. H., Campbell, B., Yang, S. 2002 AAS 34.4202
3. Frink, S., Quirrenbach, A., Fischer, D., Röser, S., Schilbach, E. 2001 PASP 113, 173
4. Frink, S., Mitchell, D. S.,Quirrenbach, A., Fischer, D.A., Marcy, G.W., Butler, R.P. 2002 ApJ 576, 478
5. Hatzes, A.P., Cochran, W.D. 1993 ApJ 413, 339
6. Hatzes, A.P., Cochran, W.D. 1994a ApJ 422, 369
7. Hatzes, A.P., Cochran, W.D. 1994b ApJ 432, 763
8. Jimenez, R., Flynn, C., Kotoneva, E. 1998 MNRAS, 299, 515
9. Marcy, G. W. & Butler, R. P. 1992, PASP, 104, 270
10. Marcy, G. W., Butler, R. P., Fischer, D. A., & Vogt, S. S. 2003, ASP Conf. Ser. 294: Scientific Frontiers in Research on Extrasolar Planets, 1
11. Mitchell, D.S., Frink, S., Quirrenbach, A., Fischer, D.A., Marcy, G.W., Butler, R.P. 2003 AAS 33, 1440
12. Shao, M. et al. 2003 - unpublished
13. Setiawan, J., Hatzes, A. P., von der Lühe, O,Pasquini, L., Naef, D., da Silva, L., Udry, S., Queloz, D., Girardi, L. 2003 A&A 398, L19
14. Walker, G.A.H., Yang, S., Campbell, B., Irvin, A.W. 1989 ApJ, 343, L21

Search for Planets Companion to Late G-type Giant Stars

Yujuan Liu*, Gang Zhao* and Yuqin Chen*

* *National Astronomical Observatories, Chinese Academy of Sciences Beijing 100012, China*

Abstract.
The searching for 'other worlds' is one of the oldest well known scientific questions. With the first discovery of planets around the pulsar PSR1257+12 (Wolszczan & Frail 1992) and the main sequence star 51 Peg (Mayor & Queloz 1995), a new field of astronomy has been opened. Most of the planets have been found around main-sequence solar-mass stars. However in this investigation, we present our plan to search for planets around intermediate-mass late G type giant stars ($1.5 - 3M_{\odot}$) with cooperation with Japanese astronomers and give the preliminary results of this research. A long-term monitoring of the Doppler velocities of stars will be carried out with mid-size telescopes at Okayama (Japan) and Xinglong (China).

INTRODUCTION

When the first extrasolar planet system was confirmed by Wolszczan & Frail with pulsar method in 1992, it really surprises to everyone, because the parent star is not a normal star in the main sequence like the sun, but rather around a pulsar, PSR 1257+12 (Wolszczan & Frail, 1992). It was not until three years later, with the discovery of the companion to 51 Pegasi (Mayor & Queloz, 1995) that the first extrasolar planetary system was found in orbit around a solar type star. And it is also the first extrasolar planetary system identified with Radial Velocity Method. There are more than 130 extrasolar planets has been found up to now.

In this paper we will briefly introduce the planet detection techniques, the reason why we search planets among late G-type giant stars, the preliminary result of this research, and our joint project with Japanese astronomers.

PLANET DETECTION TECHNIQUES

Extrosolar planets can be detected by direct and indirect methods. Each detection method is characterized by some observable characters of intrinsic physical parameters of the planet.

Direct Imaging

A planet surround a star with brightness L_* can acquire a brightness L_p by reflection, which given by:

CP752, *Stellar Astrophysics with the World's Largest Telescopes: First International Workshop*
edited by J. Mikołajewska and A. Olech

$$L_p = \frac{AL_*}{8}(\frac{R_p}{a})^2\Phi(t) \quad (1)$$

where $\Phi(t)$ is an orbital phase factor given by $\Phi(t) = 1 - \sin i \sin(2\pi t/P)$, where i is the orbit inclination with respect to sky plane.

Looking for planets orbiting around star by direct image is very hard. At optical wavelengths, the brightness ratio L_p/L_* is very small: it is $2.5 \cdot 10^{-9}$ for Jupiter, but in mid-infrared wavelengths the ratio will drop to $2.5 \cdot 10^{-6}$, making detection a little easier. Therefore, astronomers are thinking about detecting planet in mid-infrared wavelengths.

Dynamical Perturbation of The Star by The Planet

When a star has a companion planet, it will make a small circular orbit with their common center of mass as and the period naturally equal to the planet orbital period. These results in perturbations of three star's observable characters: radial velocity, angular (or astrometric) position, and time of arrival of signals.

- **Doppler Detection Method.** The wavelength of spectral lines of a star with a planet will varies periodically, because the amount of the wavelength shift corresponds to the amount of radial velocity varied which is originated from this wobbling motion. By detecting such periodic small wavelength shift of spectral lines, we could discover stars with planets. The precision in radial velocity measurements is presently 3 m/s (Butler et al. 1996), making it is easy to detect Jupiter-mass planets. But there is no hope to detect an earth-mass planet.
- **Astrometric Method.** Astrometry is used to detect the periodic wobble that a planet induces on the position of its parent star. This technique offers a great advantage: determination of an unambiguous mass and orbital inclination of a planet. The limitations to this method are the distance to the star and the positional variations of the photometric center because of the star spots. For the ground equipment, the best measurements of this method are achieved by Keck telescopes and VLTI (Very Large Telescope Interfeometer), where on reach a precision of 20 μarcsec.
- **Pulsar Timing** This method can detect planets by measuring the periodic variation in the pulsar arrival time which can detect earth-mass and even smaller planets orbiting a pulsar.

Transit Photometry

When a planet pass in front of a star along the line of sight from the observer, the luminosity of the star will changes periodically. The limitation of detect is about half of Earth size for a 1 AU orbit about a $1M_{\odot}$. Planets with orbital period more than two years are not detectable yet, since the possibility of these planets align along the line of sight to star is very small.

There are several planets have been identified since the first transit planet was found by Henry (Henry et al. 2000). This method has a great advantage: we can measure the planet's mass directly since the orbiting inclination of the planets we detected is equal to 90°.

THE PLANETS ACCOMPANY TO LATE G-TYPE GIANT STARS

Although there are more than 130 planets has been found, most of them are accompany to solar type stars, in contrast, we concentrate on finding planets around stars heavier than the Sun (i.e., $2-3M_{\odot}$). Such searches can help us to understand planet formation. Searching for planets accompany to intermediate-mass star has some special meaning, because the lifetime of surrounding disk of intermediate-mass may difference to solar-mass stars' (Haisch et al. 2001a, 2001b), and the existence of gas giant planets around such stars would constrain the timescale of formation of gas giant planet. Such survey will test the true value of the disk instability scenario, where planets can form in the time of thousands of years (Boss 1998). In the observation, massive stars on main sequence have very few and show broadened absorption lines in their spectra, so it is hard to have a precise radial velocity measurement. On the other hand, the intrinsic variability of evolved star with spectral type K or later is very large ($\geq$ 100 m/s), therefore, up to now, there are only two planetary systems have been found ie, ι Dra (Frink et al. 2002) and HD47536 (Setiawan et al. 2003). As the radial velocity of late-G giant can be determined precisely by using many sharp lines, which also show an intrinsic radial velocity around 20 m/s (Frinck at el. 2001), so late-G giant are most suitable for Doppler survey detection.

For the above reason, some Japanese astronomers began to search for planets around late-G giants using the High Dispersion Echelle Spectrograph (HIDES, Izumiura 1999) on the 1.88-m telescope at Okayama Astronomical Observatory (OAO). And following they start a joint project with National Astronomical Observatories of China.

The first planet companion to late G-type star has been identified by Sato et. al in 2003. The mass of parent star HD104985 is $1.6M_{\odot}$, and the limitation mass of the planet is $2.7M_{Jup}$. The observation data is shown in Figure 1.

Okayama 1.88-m telescope and Subaru 8-m telescope are already equipped with iodine cell filter which is used to wavelength calibration. We have also installed an iodine-cell filter for Xinglong 2.16-m telescope, which can be used to detect planetary system at the end of this year. A long-term monitoring of the Doppler velocities of stars is to be carried out with telescopes at Okayama and Xinglong.

From the Hipparcos Catalogue (ESA 1997), we select the stars which have $V < 6^m$, $0.6 < B-V < 1.0$, and $-3.0 < M_V < 2.0$. The totally about 300 stars are selected. The resolution of spectra taken from Okyama 1.88-m telescope is $R = 65000$, the wavelength coverage from 5000 – 6100 Å with S/N higher than 200. The resolution, wavelength coverage and S/N of spectra taken from Xinglong telescope are 37000, 5000 – 6100 Å, and higher than 250 respectively. The wavelength calibration is carried out by means of an iodine absorption cell, locating at the entrance of spectrometer, superposes a reference iodine spectrum directly on the stellar spectrum (Bulter 1996).

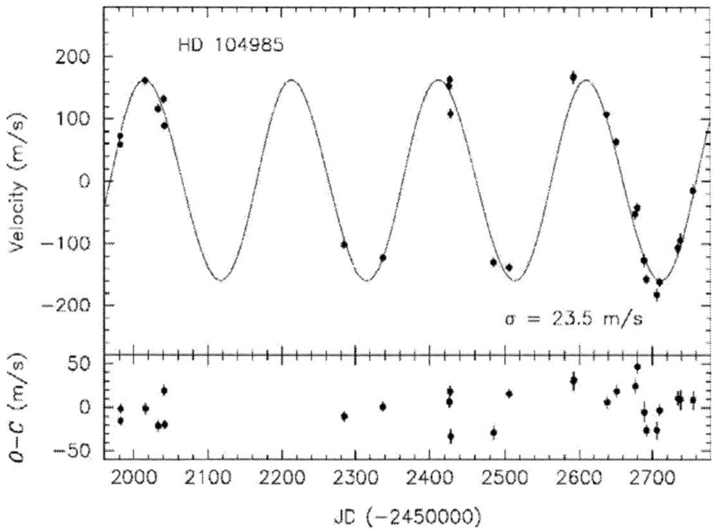

FIGURE 1. The observation data of HD104985 on radial velocity (Sato et. al. 2003). The solid line shows the Keplerian orbit fit.

We expert we can identify a few of planets through the 300 late G-type giant stars' survey.

ACKNOWLEDGMENTS

We thank our Japanese collaborator Drs. Sato and Noguchi for interesting discussion. This research is supported by the National Natural Science Foundation of China under grant number 10433010 and NKBRSF No. G1999075406.

REFERENCES

1. Boss, A. P., *ApJ*, **503**, 923 (1998)
2. Bulter, R. P., Marcy, G. W., Williams, E., McCarthy, C., Dosanjh, P., and Vogt, S. S., *PASJ*, **108**, 500 (1996)
3. Frink, S., Mitchell, D. S., Quirrenbach, A., Fischer, D., Marcy, G. W., and Bulter, R. P., *ApJ*, **576**, 478 (2002)
4. Frink, S., Quirrenbach, A., Fischer, D., RÖser, S., and Schilbach, E., *PASJ*, **113**, 173 (2001)
5. Haisch, K. E. Jr., Lada, E. A., and Lada, C. J., *AJ*, **121**, 2065 (2001a)
6. Haisch, K. E. Jr., Lada, E. A., and Lada, C. J., *ApJ*, **553**, L153 (2001b)
7. Henry, G. W., Marcy, G. W., Butler, R. P., and Vogt, S. S., *ApJL*, **529**, L41 (2000)
8. Izumiura, H., in *Proc. 4th East Asian Meeting on Astronomy*, edited by P. S. Chen, Kunming: Yunnan Observatory, 1999, p. 77
9. Mayor, M., and Queloz, D., *Nature*, **378**, 355 (1995)
10. Sato, B., Ando H., Kamber E. et al., *ApJL*, **597**, L157 (2003)
11. Setiawan, J., et al., *A&A*, **398**, L19 (2003)
12. Wolszczan, A., and Frail , D. A., *Nature*, **355**, 1451 (1992)

Physical characterization of TNOs by low resolution spectroscopy

Tomasz Kwiatkowski

Poznan Astronomical Observatory, A. Mickiewicz University, Słoneczna 36, 60-286 Poznań, Poland

Abstract. Trans-Neptunian objects (TNOs) represent a population of bodies that never approached the Sun and preserved information on the conditions in the primordial planetary nebula. Unfortunately, little is still known about their physical properties. Low resolution spectroscopy of TNOs can change this but it requires the largest, 8 – 10 m class telescopes to be used.

INTRODUCTION

Trans-Neptunian objects (hereafter TNOs) occupy the outer zone of the Solar System. They are primitive remnants from the early stages of the Solar System formation. From the dynamical point of view, TNOs can be divided into three categories: the Plutinos with orbits between 30 and 50 AU, locked in the 2:3 resonance with Neptune; the classical TNOs, also known as "Cubewanos" (after the first discovered TNO — 1992 QB1), which differ from Plutinos only by not being trapped in resonances; and the scattered disk objects (hereafter SDOs), with elongated, highly inclined orbits. The later are thought of as objects scattered out of the original TNO belt. Finally, there is a population of Centaurs on unstable orbits between those of Jupiter and Neptune. Their high orbital inclinations make them similar to SDOs and dynamical studies connect them to the main population of TNOs.

At the time of this writing there are 800 TNOs (Plutinos and Classical) as well as 147 Centaurs/SDOs known. Due to their faintness, however, very little is known about physical properties of these bodies. They contain primitive and thermally unprocessed material and that is why their study can provide the information about early stages of the evolution of the Solar System as well as the evolution of disks around young stars. TNOs are also believed to be parent bodies of the short period comets.

EVOLUTION OF TNOS

Recent theoretical and laboratory works make it possible to draw a scenario of the evolution of TNOs surfaces.

TNOs accreted at large heliocentric distances and should be built of different ices (mostly water, carbon monoxide, nitrogen and methane) mixed with rocks and dust. Their surfaces are exposed to cosmic rays, which damage molecular bonds in the icy material, producing complex, in many cases reddened, molecules [1]. This process is effective down to a 1 m depth. At the very surface, which have been exposed to the

CP752, *Stellar Astrophysics with the World's Largest Telescopes: First International Workshop*
edited by J. Mikołajewska and A. Olech

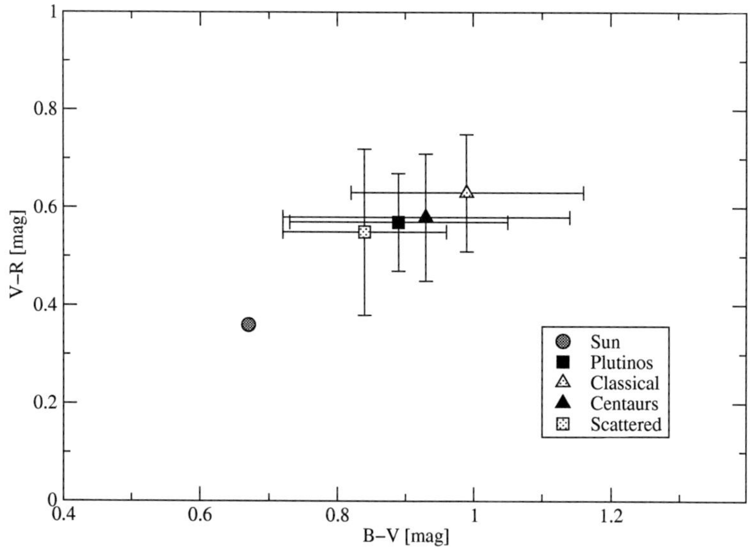

FIGURE 1. Plot of the average colors of different dynamical classes of TNOs and Centaurs, based on data from [5]. No significant differences are visible among them.

highest irradiation levels, thin crust can be formed. It contains many complex, carbon compounds which are both dark and neutral to light [2].

The crust can be further altered by mutual collisions between TNOs, which bring pristine, neutral ices to the visible surfaces [3]. Finally, for those TNOs which have not experienced collisions, micrometeorite impacts can slowly crush the dark, color neutral crust, exposing the reddened material hidden below it [4].

All these processes should work differently for dynamically "cold" TNOs (such as Classical TNOs and Plutinos, which move along the ecliptic on circular orbits, with low relative velocities) and dynamically "hot" SDOs on elongated, highly inclined paths.

PHYSICAL PROPERTIES OF TNOS

Twelve years after the discovery of the first TNO the broadband photometry was obtained for more than 100 objects and spectra are available for 23 of them.

Photometry

A color-color plot (Fig. 1) reveals no correlations between dynamical and color properties of TNOs ([5]) even though several earlier studies based on smaller datasets suggested the existence of two separate groups, one with neutral, the other with red colors.

Time-resolved photometry shows that about one-third of TNOs display light variations with peak-to-peak ranges bigger than 0.15 mag, and typical periods between 4 and 12 hours. The absence of significant color changes during rotation makes as believe the brightness variations are caused mainly by non-spherical shapes of these bodies [6].

Spectroscopy

Visible and near-infrared spectroscopy of TNOs enables more detailed studies of the reflectance properties of these objects. Typical visible spectra of TNOs and Centaurs are featureless with gradients ranging from 0 %/100 nm for gray objects to as much as 70 %/100 nm for extremely red ones (Fig. 2).

For several objects, however, broad, weak absorption bands were detected. For example, a careful analysis of the spectrum of the gray object 2003 AZ84 reveals a weak band from 500 to 800 nm (not visible in Fig. 2), which has a depth of only 3% of the continuum [9]. Similar, flat spectra are observed for the primitive, C-type asteroids, which have the same absorption band, centered at 700 nm.

Detection of water

There is a strong correlation between the presence of absorption feature at 700 nm and the infrared 3000 nm feature in phyllosilicates, the later indicating the presence of water of hydration [7]. Thus the 700 nm band can act as an indicator of the aquanous alteration of minerals in the Solar System bodies. This process, however, requires the presence of water ice as well as the energy to melt it to the liquid phase.

Water ice has been found on the number of TNOs by the direct detection of the 1500 nm and 2000 nm signatures in the infrared spectra [8]. It is still unclear what was the source of energy responsible for heating it up to the liquid phase on those surfaces, which contain hydrated minerals.

Another absorption band found in the visible spectra of 2003 AZ84 and 2002 GZ32 is centered around 430 nm and has a width of 200 nm [9]. However, these spectra are quite noisy, and the reality of this feature should still be confirmed.

Models of surfaces

There are several laws describing the light scattering by surfaces of the atmosphereless bodies—the most often used is that of Hapke [10],[11]. It is based on data obtained during laboratory experiments with mixtures of different materials as well as meteorite samples and can be used to obtain theoretical spectra, which can be compared with the observed ones.

One of the problems that arise when this theory is applied to TNOs is the lack of reliable albedos which control the vertical shift of spectra [12]. Albedos, which can be determined from direct imaging, infrared fluxes or dynamical studies of binary systems,

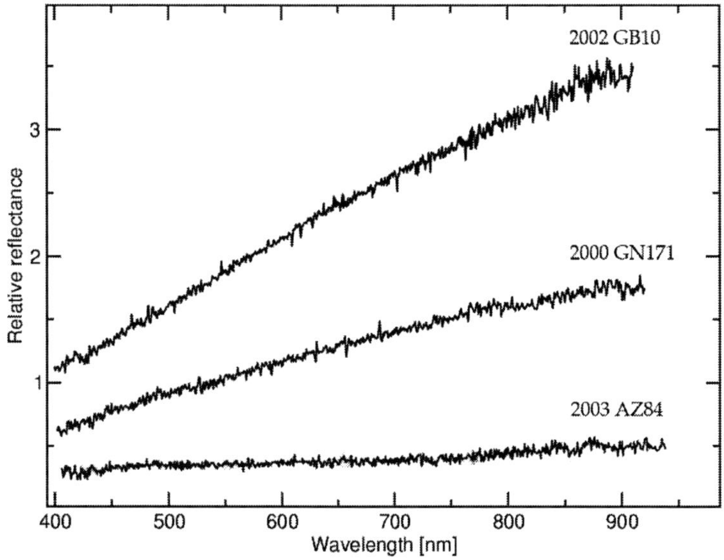

FIGURE 2. Visible spectra of two TNOs (2003 AZ84 and 200 GN171) and one Centaur (2002 GB10), obtained during ESO Large Program [9]. All graphs were shifted vertically for better visibility.

are known only for about 20 TNOs and Centaurs. For the rest of objects a standard value of 0.04 is assumed.

Up till now several models of TNOs surfaces has been derived [13]. They usually use Titan tholins or kerogen-type materials to obtain the red slope of the observed spectra and mix them with carbon-rich minerals to lower the albedo. Different ices are added to represent absorption features which, when properly accounted for, can help to obtain a unique solution.

More spectra of TNOs needed

From the above discussion it is clear new spectroscopic observations are needed to improve our understanding of TNOs surfaces. The faintness of these objects (20-24 mag) requires the largest 8-10 m class telescopes to be used.

There is no special demand for a high signal-to-noise ratio and usually S/N=20 is acceptable. It is often possible to combine pixels trading resolution for S/N as the expected absorption features are very wide and the spectrograph resolution of R=100-200 is sufficient. The problems with detection of very weak absorption bands mentioned earlier were mainly caused by difficulties with removal of telluric lines as well as the solar features.

On the celestial sphere TNOs are spread along the ecliptic band which makes them easily accessible for observations. For southern telescopes there is a slight preference for March to September observing period, because then TNOs culminate at higher elevations then during the rest of the year.

It should be remembered that TNOs move by 2-3 arc seconds during an hour so it is better to combine several 15-30 min exposures than to risk a single, 60 min observation. Apart from the normal, spectrophotometric standards, solar analog stars should also be observed on the same night to remove the solar flux from the final spectra.

REFERENCES

1. Moore, M. H., Hudson, R. L., and Ferrante, R. F., *Earth Moon and Planets*, **92**, 291–306 (2003).
2. Moroz, L. V., Baratta, G., Distefano, E., Strazzulla, G., Starukhina, L. V., Dotto, E., and Barucci, M. A., *Earth Moon and Planets*, **92**, 279–289 (2003).
3. Luu, J., and Jewitt, D., *Astron. J.*, **112**, 2310–2318 (1996).
4. Cooper, J. F., Christian, E. R., Richardson, J. D., and Wang, C., *Earth Moon and Planets*, **92**, 261–277 (2003).
5. Hainaut, O. R., and Delsanti, A. C., *Astron. Astrophys.*, **389**, 641–664 (2002).
6. Sheppard, S. S., and Jewitt, D. C., *Astron. J.*, **124**, 1757–1775 (2002).
7. Vilas, F., *Icarus*, **111**, 456–467 (1994).
8. Fornasier, S., Dotto, E., Barucci, M. A., and Barbieri, C., *Astron. Astrophys.*, **422**, L43–L46 (2004).
9. Fornasier, S., Doressoundiram, A., Tozzi, G. P., Barucci, M. A., Boehnhardt, H., de Bergh, C., Delsanti, A., Davies, J., and Dotto, E., *Astron. Astrophys.*, **421**, 353–363 (2004).
10. Hapke, B., *J. Geophys. Res.*, **86**, 3039–3054 (1981).
11. Hapke, B., "Combined Theory of Reflectance and Emittance Spectroscopy," in *Remote Geochemical Analysis: Elemental and Mineralogical Composition*, edited by C. Pieters and P. Englert, Cambridge Univ. Press, New York, 1993, pp. 31–42.
12. Grundy, W. M., and Stansberry, J. A., *Earth Moon and Planets*, **92**, 331–336 (2003).
13. de Bergh, C., Boehnhardt, H., Barucci, M. A., Lazzarin, M., Fornasier, S., Romon-Martin, J., Tozzi, G. P., Doressoundiram, A., and Dotto, E., *Astron. Astrophys.*, **416**, 791–798 (2004).

HIGH PRECISION TECHNIQUES

Stellar Speckle Interferometry with Large Telescopes

Yuri Balega

Special Astrophysical Observatory, Nizhnij Arkhyz 2/72, 369167 Zelenchuk region, Russia

Abstract.
Speckle interferometry is a method for attainment high angular resolution using full aperture of the large telescope. We give a brief description of the main principles of the method and the history of its development at large telescopes. The main fields of application of the method are the binary and multiple star research, study of the circumstellar gas and dust structures both for young objects and for stars at the latest stages of their evolution. Stellar diameters of nearby supergiants can be measured in different bands using the largest world telescopes. We discuss the perspectives of the method adaptation at the telescopes with segmented mirrors.

CP752, *Stellar Astrophysics with the World's Largest Telescopes: First International Workshop*
edited by J. Mikołajewska and A. Olech

Spectropolarimetry of single and binary stars

Tim J Harries

School of Physics, University of Exeter, Stocker Road, Exeter EX4 4QL.

Abstract.
Spectropolarimetry is a photon-hungry technique that will reach fruition in the 8-m telescope age. Here I summarize some of the stellar spectropolarimetric research that my collaborators and I have undertaken, with particular emphasis on the circumstellar environment of massive stars, symbiotic binaries, and star formation.

INTRODUCTION

Stellar spectropolarimetry[1] is a relatively underused technique that has enormous diagnostic potential, particularly for research into circumstellar matter and close binary systems. The principle difficulty with the method is that very high signal-to-noise observations are required, since typical astrophysical polarimetric signatures may be on the order of a few tenths of a percent. When photon noise is dominant, the polarimetric precision ΔP goes as

$$\Delta P \approx \frac{100}{\sqrt{N}} \tag{1}$$

where ΔP is in percent and N is the number of photons. One can see that a S/N of 1000 or more per pixel is needed to measure a spectropolarimetric signal at the 0.1% level. At moderate resolutions ($R \sim 5000$) this limits even 4-m class telescopes to relatively bright objects (say $V < 12$). Higher dispersions and fainter objects necessitate even larger apertures, and fortunately spectropolarimetric instrumentation is present or planned on most of the world's 8-m facilities.

The origin of linear polarization in a stellar context is usually dichroic absorption by dust, or scattering by dust, molecules or electrons (although linearly polarized thermal emission from aligned grains is sometimes observed in the millimetre regime). For an unresolved source an asymmetric geometry with respect to the observer's sightline is required, for example the alignment of grains by a magnetic field, an asymmetric circumstellar distribution of material such as a disc, or intra-binary scattering. Cancellation between polarized radiation from different regions often reduces the observed polarization to the level of a few percent – nonetheless the polarimetry enables one to probe geometrical parameters of an unresolved object.

[1] Note that this paper focuses on the use of linear spectropolarimetry, although of course circular spectropolarimetry is an extremely powerful tool for probing stellar magnetic field structures, both at the stellar surface and in circumstellar material (e.g. [1, 2]).

CP752, *Stellar Astrophysics with the World's Largest Telescopes: First International Workshop*
edited by J. Mikołajewska and A. Olech

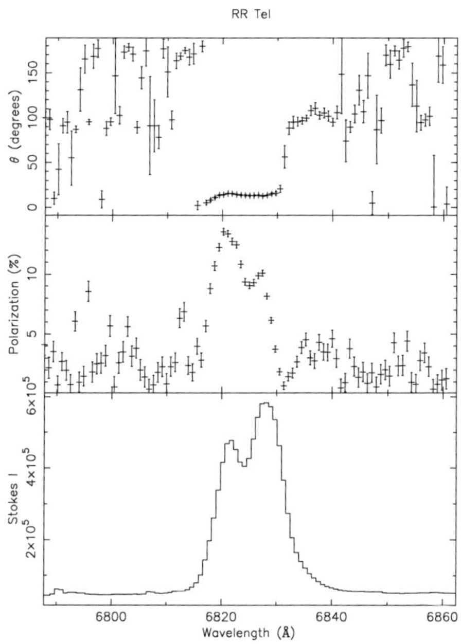

FIGURE 1. Spectropolarimetry of the 6825Å Raman-scattered line of D-type symbiotic RR Tel. This 'triplot' shows the intensity spectrum (bottom panel), the polarization magnitude in percent (middle panel), and the position angle of the polarization (top panel).

In this paper I present a synopsis of some of the stellar spectropolarimetric studies that my collaborators and I have undertaken. I concentrate on three broad areas: Raman-scattering in symbiotic binaries, and its use in measuring orbits in these long-period systems; the use of recombination-line polarimetry to measure global structure in the winds of massive stars; and the use of spectropolarimetric in examining the circumstellar structure of pre-main-sequence stars. In the final section I give a brief account of the instruments and likely avenues of research in the 8-m telescope age.

RAMAN SCATTERING IN SYMBIOTIC BINARIES

The spectra of symbiotic stars exhibit both molecular absorption bands and nebular emission lines. It is now known that this so-called 'combination spectrum' is formed due to binary interaction between a mass-losing red-giant star and a hot companion (typically a white dwarf), which ionizes the cool wind from the giant. Symbiotic stars are thought to be detached systems [3], with accretion onto the hot component occurring from the wind rather than via Roche-lobe overflow. Systems that show a stellar spectrum in the IR are classified as S-type systems, while those that show dust emission are denoted D-type.

In a seminal paper Schmid [4] determined that the $\lambda\lambda 6825, 7082$Å emission lines that are observed in $\sim$50% of symbiotic systems arise from Raman scattering of the

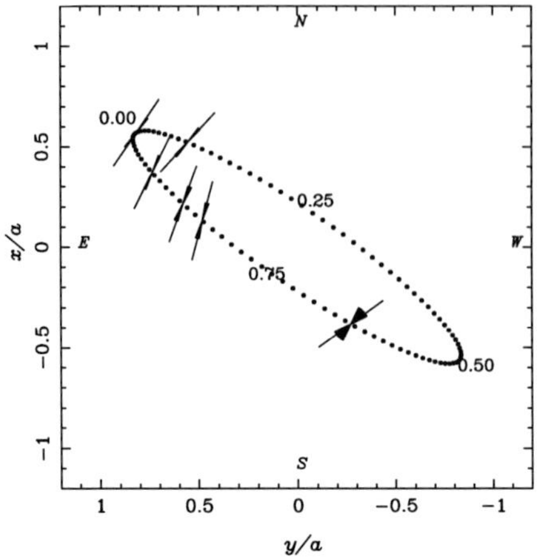

FIGURE 2. The polarimetric orbit of SY Mus. The O VI source is located at $(0,0)$ and the filled dots show the position of the red giant every $0.01P$. Short thick 'bowties' indicate the measured PAs and their uncertainties, the long thin lines show the calculated PAs at the times of the observations. The spectroscopic phases of quadrature and conjunction are also marked.

$\lambda\lambda 1032, 1038$Å O VI resonance doublet by neutral hydrogen. Raman scattering is the inelastic analogue of Rayleigh scattering: the frequency of the scattered photon is different to that of the incident photon, and the scatterer is left in an altered quantum mechanical state. In this case the O VI photons scatter in the far-red wing of Lyβ and the scattering H^0 atom is left in the $n = 2$ state. By conservation of energy the frequency of the Raman photon, ν_r, is given by

$$\nu_r = \nu_P - \nu_{Ly\alpha} \tag{2}$$

where $\nu_{Ly\alpha}$ is the frequency of the Lyα transition and ν_P is the frequency of the parent photon.

The asymmetry of the scattering geometry, with O VI photons produced near the hot star scattering in the cool star's wind, coupled with the dipole nature of the scattering event, leads naturally to linear polarization [5]. Extensive surveys of symbiotic systems ([6, 7]) revealed that the Raman lines often have a complex structure, typically comprising two or three peaks in both intensity and polarization accompanied by a 90° position angle (PA) flip and/or PA rotation (see, e.g., Figure 1). The symbiotics show large line polarizations (up to $\sim 15\%$), with the strongest and most highly polarized lines found in D-type (dusty) Mira systems.

Numerical modelling of the Raman-line formation process ([8, 9, 10]) indicates that the PA of the blue wing of the Raman-line polarization is perpendicular to the plane containing the binary line-of-centres and the observer's line of sight. The models further demonstrate that the magnitude of the line polarization (P_l) is also a simple function of orbital phase

$$P_l = A \sin^2 \alpha \tag{3}$$

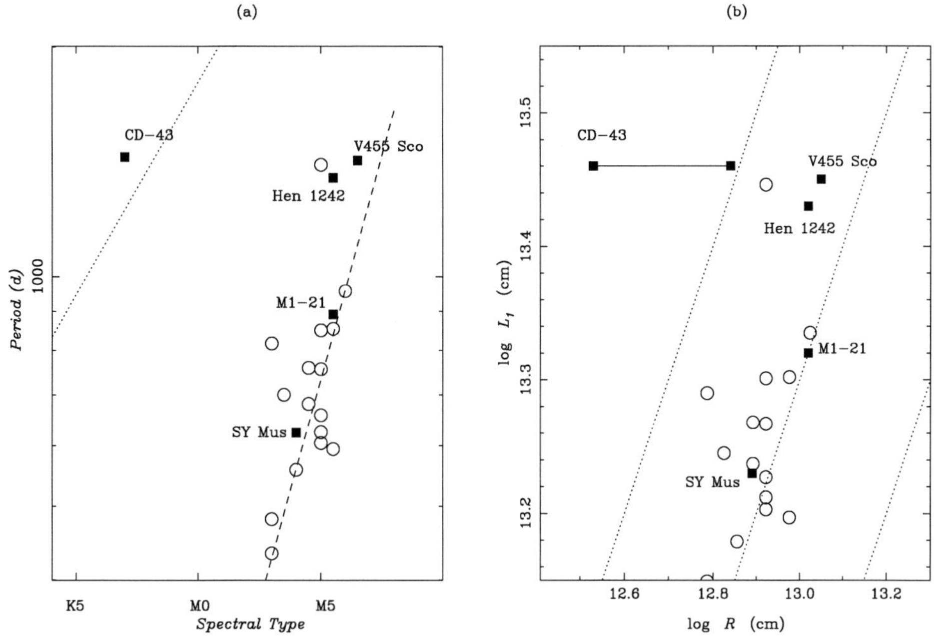

FIGURE 3. Plots of *(a)* orbital period against spectral type, and *(b)* the distance to the L_1 point against stellar radius. The targets from this study are plotted as filled squares, and the open circles correspond to other symbiotics of 'known' period listed in [3]. The dashed line in *(a)* corresponds to the minimum period relationship while the dotted line connects the longest period systems. The dotted lines in panel *(b)* correspond to $R = L_1$, $R = 2L_1$, and $R = 3L_1$. The solid line joins the maximum radius of CD−43°14304 ($100R_\odot$) to the radius predicted from its spectral type ($48R_\odot$).

where α is the angle between the observer's sightline and the binary line-of-centres, and A is a scale factor which accounts for geometrical and physical parameters (interbinary distance, stellar radii; mass-loss rate, velocity law, etc). By observing the change in line polarization with time, it is possible to derive orbital parameters (e.g., period, inclination, PA of line-of-nodes) with analogous methods to those used for visual binaries. The first 'spectropolarimetric orbit', that of SY Mus (Figure 2; [11]), was followed by those of AG Dra and Z And ([12, 13]), while follow-up observations yielded orbits of four additional systems (CD−43°14304, Hen 1242, M1-21, V455 Sco) [14], thereby doubling the number of known orbits with $P > 1000$d.

It has been shown by Mürset & Schmid ([3]) that there is a strong correlation between orbital period and spectral type of the cool component, with the majority of systems lying close to a straight line relating the spectral subtype to the orbital period. For a given subtype no systems are found with substantially shorter periods. The minimum period locus is given by

$$\log P_{\rm min} = 0.117S + 2.28 \tag{4}$$

where S is the M-class subtype and $P_{\rm min}$ is in days (see Figure 3*a*). Mürset & Schmid proposed that this period limit is related to the red giant radius, which was found to be

$L_1/2$ for most systems (where L_1 is the distance from the primary to the inner Lagrangian point). Essentially this means that for binary to appear as a symbiotic it must be close enough for significant wind accretion onto the hot component, but wide enough that the binary is detached (Figure 3*b*). Our systems with spectropolarimetric orbits conform to this relation, excepting CD−43, which has an exceptionally early spectral-type for its cool component (K7).

The spectropolarimetric approach to orbit determination has unique advantages. First, because the observed quantities can be determined to a small fraction of the orbital amplitudes (e.g., PA to $\sim 1^\circ$ in 180°), the approach is extremely efficient – e.g., in the favourable case of SY Mus, we were able to determine correct orbital characteristics from just four observations, whereas the same information from traditional radial-velocity techniques required many observing nights [15]. Secondly, the amplitude of variation of the polarization is *independent* of orbital period, so that adequately sampled datasets are capable of yielding accurate orbital characteristics for systems with periods of from a few months to many years (unlike normal spectroscopic methods). Our group, along with Schmid's at ETH Zürich, are continuing to obtain data on various symbiotics with the aim of establishing periods for many more S-type systems, and obtaining the first reliable D-type orbit. The latter goal is obviously long-term, since D-type systems are likely to have periods of many decades!

THE GEOMETRY OF STELLAR WINDS

Massive stars shed mass via high-speed, radiation-driven winds. Mass-loss rates of 10^{-7} to 10^{-4} $M_\odot$ yr^{-1} are typical, with terminal speeds of 1000–3000 km s^{-1}. Stellar wind models are often used to fit recombination line profiles (such as Hα) in order to determine mass-loss rates. This approach essentially reduces to 'counting' recombinations, based on the standard assumptions of homogeneity and sphericity. Departures from either will lead to an overestimate of the true mass-loss rate: a wind that is clumpy on small scales, or is global distorted, will have an increased rate of recombinations (which is a density-squared process) per unit mass, so a structured wind will produce greater line emission than a smooth wind with the same mass-loss rate.

So how applicable is the assumption of sphericity? The winds cannot be imaged directly, but fortunately polarimetry provides an indirect probe of the geometry of the wind near to the star. Continuum photons from the star will pass through the ionized wind, and some will be electron-scattered which introduces linear polarization to the starlight. The nett polarization will be zero for a spherical distribution of electrons, but will be non-zero for any geometry that is not circularly symmetric with respect to the observer's line of sight. The recombination line emission, which is produced further out in the wind, will 'see' a lower electron-scattering column, and will therefore be less polarized than the continuum. Spectropolarimetry of a non-spherical wind should therefore show a polarized continuum (on the order of a few tenths of a percent) with the magnitude of the polarization dropping at the emission line wavelengths, as the polarized continuum is diluted by unpolarized line emission. This known is the 'line effect'. As an illustration I have plotted polarization spectra of two Wolf-Rayet (W-R) stars in Figure 4: note the obvious decline in continuum polarization magnitude across the emission lines,

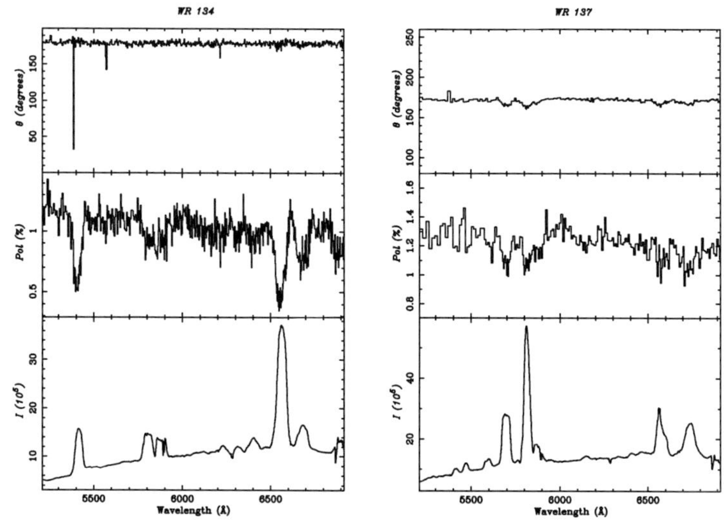

FIGURE 4. Spectropolarimetry of the WN star WR 134 and the WC star WR 137, binned to a constant polarization of 0.1% and 0.05% respectively. Note the strong depolarizations at the emission line wavelengths.

a sure indication that the stellar winds in these objects depart from spherical symmetry.

For WR137 we have other indicators that wind may be asymmetric. This WC star has an O-star binary companion in an eccentric orbit with a period of ~ 12 years. At periastron the interaction of the winds from the two components provides the necessary conditions for the condensation of warm carbonaceous dust grains, which produce a near-IR excess [16]. There is an intimate relationship between the solar-radius-scale geometry of the stellar wind from polarization, and the larger scale distribution of newly formed dust (many hundreds of AU) – see Figure 5 and [17]. The link between near-star geometry and a large-scale nebula has also been established for the Luminous Blue Variable (LBV) AG Car [18]. Spectropolarimetry of this object shows a strong line effect at Hα, along with substantial variation ($> 1\%$) in continuum polarization. The PA of the intrinsic polarization is aligned with the major axis of the AG Car ring nebulae and perpendicular to its jet, indicating that the resolved circumstellar environment is already present within a few stellar radii of the central object.

The magnitude of the continuum polarization (P_c) for a given stellar wind geometry and viewing angle (i) may, in the optically thin limit, be computed analytically [20]:

$$P_c \approx 2\overline{\tau}(1-3\gamma)\sin^2 i \tag{5}$$

where $\overline{\tau}$ is the angle-averaged electron-scattering optical depth of the envelope and γ is a shape factor. The degeneracy between the stellar wind geometry and the viewing angle means that studying objects on an individual basis is rather difficult, but if a large enough sample is obtained it is possible to disentangle the two statistically. We gathered a spectropolarimetric data set of 16 W-R stars using the 4-m William Herschel Telescope (WHT) and the ISIS spectrograph [21]. A statistical analysis of these data using Monte-

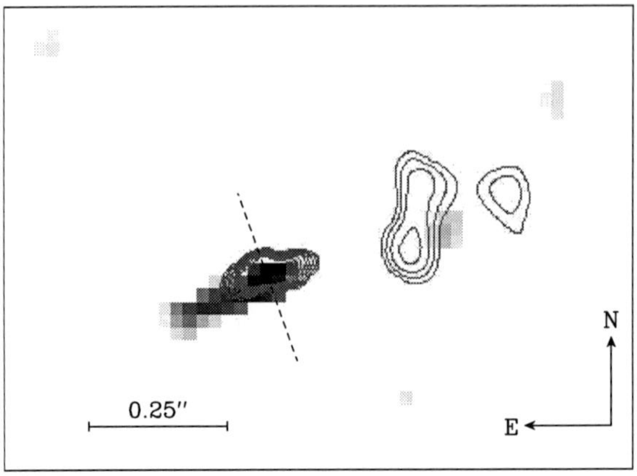

FIGURE 5. The 1998 K' maximum entropy restored images of WR137 (logarithmic greyscale), and the 1997 image (contours) adapted from Fig. 3 of [19]. The dashed line illustrates the PA of the intrinsic polarization.

Carlo based K-S tests enabled us to show that about 15% of W-R stars show a global departure from spherical symmetry. The level of the intrinsic polarization is consistent with an equator-to-pole density ratio of 2–3. We postulated that the line-effect systems represented the most rapidly rotating W-R stars, in which the rotation was having a significant affect on the wind dynamics.

Interestingly the line-effect systems are clustered towards high mass-loss rates and high-luminosity (Figure 6). Although at first sight this appears to suggest that this may be the root cause for the wind asymmetry, one should note that these quantities are very dependent on small and large-scale structure, as noted above. In other words the high mass-loss rates and luminosities determined by the models may be a result of applying spherically symmetric models. Mass-loss rates derived from thermal radio emission, which are the least model-dependent, show good agreement with those derived from the optical recombination lines. This does not mean that the mass-loss rate values are correct in an absolute sense, since both the radio continuum and the recombination lines scale as density-squared, but the agreement does indicate that the degree of wind asymmetry is similar at the radii for optical depth ~ 1 of both the optical $\sim 10^1 R_*$ and the radio $\sim 10^3 R_*$.

Since departures from spherical symmetry appear to be relatively common for the W-R stars, and also occur in LBVs, it seemed natural to search for evidence for global structure in the winds of O-supergiants. We obtained Hα spectropolarimetry of a sample of 20 Northern, and combined these data with K-band spectropolarimetry obtained with the 4-m United Kingdom Infrared Telescopes and low resolution optical spectropolarimetry from Pine Bluff Observatory [22]. Somewhat surprisingly, the spectropolarimetry revealed that the single O-supergiants typically have intrinsic polarizations of $< 0.1\%$ (which corresponds to an equator:pole asymmetry of approximately less than 1.25). It appears that the wind line-driving mechanism in rapidly-rotating O supergiants (whose winds are optically thin) is relatively insensitive to rotation compared to the optically-

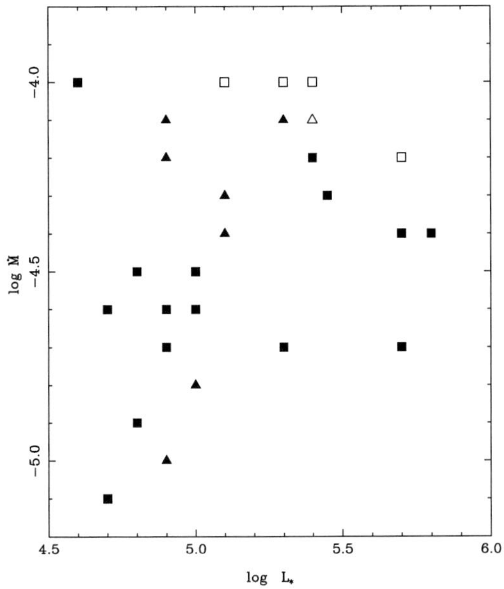

FIGURE 6. Wolf-Rayet stars plotted in the mass-loss/luminosity plane. WN stars are plotted as squares, while WC stars are plotted as triangles. Stars with a line effect are displayed as open symbols, stars with no detectable line effect are shown as filled symbols.

thick line-driving mechanism of W-R outflows.

W-R stars are also known to have small-scale clumps that appear to accelerate slowly from the star (e.g. [23]), and there is spectroscopic evidence for similar structures in the O supergiant ζ Puppis [24]. These migrating density structures could potentially be identified polarimetrically, but numerical models of density enhancements propagating through the wind predict polarizations of close to 0.1%–0.2% [25], which is uncomfortably close to the current limits on the absolute precision of polarimetric observations. It is clear that very precise techniques will be needed to probe these clumps in the winds of O supergiants.

PRE-MAIN-SEQUENCE STARS

The broad picture of low-mass star formation, from the collapse of a gravitationally unstable molecular cloud, through to a pre-main-sequence star surrounded by a dusty accretion disc, is well-established. During the final, classical T Tauri star (CTTS) phase magnetospheric accretion is thought to occur, with the star's magnetic field disrupting the disc and channelling the material onto the surface, where its kinetic energy is liberated at hot spots. However, the situation for intermediate and high-mass star formation is much more confused, with a lack of clear-cut evidence for circumstellar discs. In the pre-

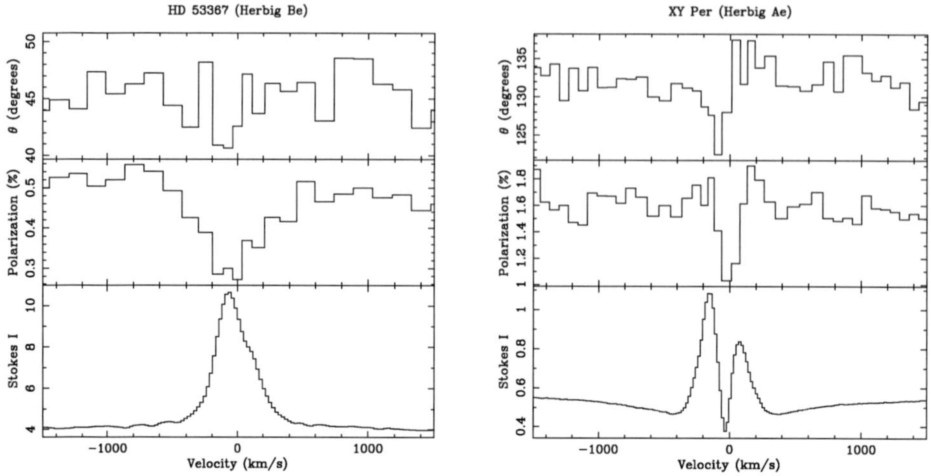

FIGURE 7. Hα spectropolarimetry of a Herbig Be star (HD 53367) and a Herbig Ae (XY Per), binned to a constant error in polarization of 0.03% and 0.1% respectively. Note that the Herbig Be star shows a classical line depolarization, whereas the polarization spectrum of the Herbig Ae star is more complex.

vious sections I have shown that line spectropolarimetry can be a powerful probe of the circumstellar geometry, and we are now in the process of applying spectropolarimetric techniques to pre-main-sequence (PMS) stars.

We have obtained a large sample of Hα polarization spectra Herbig Ae/Be stars [26]. The two classes of star show a dichotomy in spectropolarimetric characteristics. The Herbig Be stars typically show the classical depolarization line effect, which is indicative of a large volume of Hα emitting material embedded in an asymmetric envelope. The Herbig Ae stars on the other hand show a more complex polarization through the line, with some stars showing significant polarized Hα emission. This suggests that Herbig Ae stars have compact sources of Hα emission, possibly magnetospheric accretion streams, with Hα photons scattering in the surrounding asymmetric medium. It is not completely clear whether the scatterers here are dust or electrons, although thermal motions of electrons should lead to significant smearing of the line polarimetry, which is not observed. Our results indicate that the magnetospheric accretion paradigm for low-mass star formation may extend to higher masses.

A significant investment has been made in photopolarimetric studies of classical T Tauri stars (CTTS), which may show continuum polarizations of a few percent e.g. [27, 28, 29, 30]. This polarization is variable, perhaps due to changes in the illumination of the scattering envelope due to hot spots [31]. Since it was thought that the line and continuum photons from the central object were scattered by the same dusty envelope the expectation was that there should be no line effect. However our recent observations of RY Tau (Figure 8, [32]) show a dramatic change in the polarization through Hα. It appears that RY Tau is not exceptional in this regard: in our recent survey 9 out of 10 CTTS showed a detectable change in polarization across Hα [33]. These observations are qualitatively consistent with scattering of a compact source of Hα photons off a

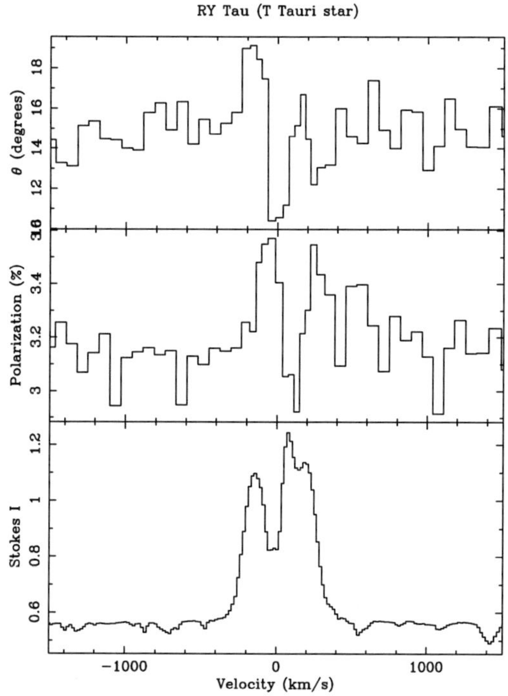

FIGURE 8. Hα spectropolarimetry of the classical T Tauri star RY Tau, binned to a constant error in polarization of 0.1%.

circumstellar disc.

A more detailed interpretation of the spectropolarimetry requires numerical modelling. As a first attempt we have examined the polarimetric signature of dipolar scattering of a compact line source off optically-thin and optically-thick dusty discs [34] using the Monte-Carlo radiative-transfer code TORUS [25]. We find that the PA dependence of the line profile is an indicator on the extent of the disc, with a double PA rotation through the line corresponding to a disc with a negligible inner hole, and a single PA rotation indicating a substantial evacuated area around the central star. By interpreting the observations in the context of these numerical simulations we will begin to place strong constraints on the circumstellar environment of PMS across the mass spectrum.

THE FUTURE

Spectropolarimetric capabilities are already available on 8-m class telescopes. For example, VLT/FORS offers both circular and linear spectropolarimetry, but at $R \sim 2500$ its maximum resolution is insufficient for most of the line polarization work described here (although the broad emission lines in supernovae can be resolved). Similarly the Keck/LRIS combination is limited essentially to continuum studies and is used predominantly for extragalactic research. The two 8-m Gemini telescopes are due to be fitted with facility polarization modulator units (GPOL), which will eventually provide a number of their optical and near-IR instruments, including a high-dispersion echelle spectrograph, with a polarimetric capability. The Prime Focus Imaging Spectrograph (PFIS) on the 10-m South African Large Telescope will provide a medium resolution ($R \sim 5000$) spectropolarimetric capability, and is ideally suited to stellar work.

Pleasingly it appears that spectropolarimetry is now viewed as a standard mode by spectrograph developers. The proliferation of spectropolarimetry in the 8-m telescope age will lead to exciting new advances in stellar astrophysics, for example in the investigation of the link between rotation and outflow asymmetries and structure in massive stars and asymmetries in core collapse supernovae, which can show a line effect ([35, 36]), and may have viewing-angle-dependent observed luminosities. Rotation and asymmetry obviously play a key role in the collapsar model for gamma ray burst progenitors [37], and high-resolution linear spectropolarimetry of the burst afterglows should yield valuable insights. Extending the spectropolarimetric study of W-R stars into the low-metallicity environment of the Magellanic clouds would provide a good test of line-driving theories. The move to longer wavelengths will also lead to new discoveries, for example by probing the circumstellar environments of highly obscured objects, such as Class I protostars that are still embedded in their nascent molecular clouds.

ACKNOWLEDGMENTS

I am grateful to the SOC for the invitation and for hosting a pleasant and interesting meeting. I am of course indebted to my research collaborators in polarimetry: Ian Howarth (UCL, UK), Ron Hilditch (St.Andrews, UK) John Hillier (Pittsburgh, USA), Regina Schulte-Ladbeck (Pittsburgh, USA), Janet Drew (Imperial College, UK), Jorick Vink (Imperial College, UK), René Oudmaijer (Leeds, UK), Ryuichi Kurosawa (Exeter, UK), Neil Symington (Exeter, UK), Bruce Babler (Wisconsin, USA), and J-F Donati (Toulouse, France).

REFERENCES

1. J.-F. Donati, G. A. Wade, J. Babel, H. f. Henrichs, J. A. de Jong, and T. J. Harries, *MNRAS*, **326**, 1265–1278 (2001).
2. M. Jardine, A. Collier Cameron, and J.-F. Donati, *MNRAS*, **333**, 339–346 (2002).
3. U. Mürset, and H. M. Schmid, *A&AS*, **137**, 473–493 (1999).
4. H. M. Schmid, *A&A*, **211**, L31–L34 (1989).
5. H. M. Schmid, and H. Schild, *A&A*, **236**, L13–L16 (1990).
6. H. M. Schmid, and H. Schild, *A&A*, **281**, 145–160 (1994).
7. T. J. Harries, and I. D. Howarth, *A&AS*, **119**, 61–90 (1996).
8. H. M. Schmid, *MNRAS*, **282**, 511–529 (1996).
9. T. J. Harries, and I. D. Howarth, *A&AS*, **121**, 15–44 (1997).
10. K. W. Lee, and H. Lee, *MNRAS*, **292**, 573–+ (1997).
11. T. J. Harries, and I. D. Howarth, *A&A*, **310**, 235–238 (1996).
12. H. M. Schmid, and H. Schild, *A&A*, **321**, 791–802 (1997).
13. H. M. Schmid, and H. Schild, *A&A*, **327**, 219–223 (1997).
14. T. J. Harries, and I. D. Howarth, *A&A*, **361**, 139–152 (2000).

15. W. Schmutz, H. Schild, U. Muerset, and H. M. Schmid, *A&A*, **288**, 819–828 (1994).
16. P. M. Williams, M. R. Kidger, K. A. van der Hucht, P. W. Morris, M. Tapia, M. Perinotto, L. Morbidelli, A. Fitzsimmons, D. M. Anthony, J. J. Caldwell, A. Alonso, and V. Wild, *MNRAS*, **324**, 156–166 (2001).
17. T. J. Harries, B. L. Babler, and G. K. Fox, *A&A*, **361**, 273–282 (2000).
18. R. E. Schulte-Ladbeck, G. C. Clayton, D. J. Hillier, T. J. Harries, and I. D. Howarth, *ApJ*, **429**, 846–856 (1994).
19. S. V. Marchenko, A. F. J. Moffat, and Y. Grosdidier, *ApJ*, **522**, 433–439 (1999).
20. J. C. Brown, and I. S. McLean, *A&A*, **57**, 141–+ (1977).
21. T. J. Harries, D. J. Hillier, and I. D. Howarth, *MNRAS*, **296**, 1072–1088 (1998).
22. T. J. Harries, I. D. Howarth, and C. J. Evans, *MNRAS*, **337**, 341–355 (2002).
23. S. Lepine, A. F. J. Moffat, and R. N. Henriksen, *ApJ*, **466**, 392–+ (1996).
24. T. Eversberg, S. Lepine, and A. F. J. Moffat, *ApJ*, **494**, 799–+ (1998).
25. T. J. Harries, *MNRAS*, **315**, 722–734 (2000).
26. J. S. Vink, J. E. Drew, T. J. Harries, and R. D. Oudmaijer, *MNRAS*, **337**, 356–368 (2002).
27. P. Bastien, *A&AS*, **48**, 153–164 (1982).
28. P. Bastien, *ApJS*, **59**, 277–291 (1985).
29. F. Menard, and P. Bastien, *AJ*, **103**, 564–572 (1992).
30. E. Gullbring, and G. F. Gahm, *A&A*, **308**, 821–827 (1996).
31. K. Wood, and B. Whitney, *ApJL*, **506**, L43–L45 (1998).
32. J. S. Vink, J. E. Drew, T. J. Harries, R. D. Oudmaijer, and Y. C. Unruh, *A&A*, **406**, 703–707 (2003).
33. J. Vink, T. Harries, J. Drew, R. Oudmaijer, and Y. Unruh, *MNRAS*, **submitted** (2005).
34. J. Vink, T. Harries, and J. Drew, *A&A*, **submitted** (2005).
35. M. Cropper, J. Bailey, J. McCowage, R. D. Cannon, and W. J. Couch, *MNRAS*, **231**, 695–722 (1988).
36. L. Wang, D. Baade, P. Höflich, and J. C. Wheeler, *ApJ*, **592**, 457–466 (2003).
37. A. I. MacFadyen, and S. E. Woosley, *ApJ*, **524**, 262–289 (1999).

The evolutionary phase of B[e] supergiants and unclassified B[e] stars

Michaela Kraus

Astronomical Institute, Utrecht University, Princetonplein 5, 3584 CC Utrecht, The Netherlands

Abstract. We present two classes of stars with yet unknown evolutionary phase: the B[e] supergiants and the so-called unclassified B[e] stars. While the B[e] supergiants are luminous post-main sequence stars with high mass progenitors, not much is known of the unclassified stars. We discuss how it might be possible to determine their evolutionary phases in the area of new large telescopes.

INTRODUCTION

Stars showing the B[e] phenomenon have several characteristics of their optical spectrum in common: they are of spectral type B, have strong Balmer emission lines, and many permitted and forbidden emission lines of predominantly low ionization metals like e.g. FeII and OI. In addition, these stars are known to possess a strong near- or mid-infrared excess due to hot circumstellar dust. Since most of the characteristics are based purely on the optical spectrum, these stars can be of quite different evolutionary phase, and Lamers et al. [1] grouped all stars with the B[e] phenomenon accordingly. The different classes they found are: B[e] supergiants, Herbig Ae/B[e] stars, and compact planetary nebula B[e] stars. The biggest group, however, is formed by stars of unknown evolutionary phase, the so-called unclassified B[e] stars. In this paper we want to concentrate on two classes of B[e] stars: the supergiants and the unclassified stars. Both classes have in common, that their evolutionary phase is still unknown and we want to strengthen the need of large telescopes for the determination of the evolutionary phases of these stars.

THE B[E] SUPERGIANTS

B[e] supergiants are luminous ($4 \leq \log L/L_{\odot} \leq 6$) post-main sequence objects. Their progenitors must have been massive ($7 < M/M_{\odot} < 85$) and probably rapidly rotating stars. The best known B[e] supergiants are located in the Magellanic Clouds. In the Milky Way only some candidates have been found. Here, due to large interstellar extinction the proper determination of distances and therefore luminosities is difficult. The optical spectra of B[e] supergiants show a hybrid character, i.e. a co-existence of very broad Balmer lines and extremely narrow forbidden and permitted low ionization metal lines. This hybrid character has led to the interpretation of two different winds: a normal line-driven (CAK-type) fast wind of low density in polar direction and a slow, high

CP752, *Stellar Astrophysics with the World's Largest Telescopes: First International Workshop*
edited by J. Mikołajewska and A. Olech

density disk-forming wind in equatorial direction where the hot dust is located (see Zickgraf et al. [2]). Kraus & Lamers [3] performed ionization structure calculations in such non-spherical winds of B[e] supergiants using a latitude dependent mass flux that increases from pole to equator. They found that with such a model hydrogen can recombine close to the stellar surface in the equatorial region leading to a hydrogen neutral disk-like structure around these luminous stars. High-resolution optical spectra of B[e] supergiants revealed strong emission of OI forbidden lines (see e.g. Kraus & Borges Fernandes [4]) which indicates the existence of a huge amount of neutral hydrogen since H and O have the same ionization potential. The additional detection of CO band emission in the near-infrared in several B[e] supergiants (e.g. McGregor et al. [5]; [6]; [7]) completes the picture of the circumstellar disk leading to the following schematic representation of its temperature structure (from inside out): [OI] emission ($10\,000\,\mathrm{K} \leq T \leq 6\,000$) – CO band emission ($5\,000\,\mathrm{K} \leq T \leq 2\,000$) – dust emission ($T \leq 2\,000\,\mathrm{K}$).

In the HR diagram, B[e] supergiants share their location with other massive post-main sequence stars like Wolf-Rayet stars and Luminous Blue Variables (LBVs). But the connection of B[e] supergiants with these other evolutionary states is still an unsolved problem and even newest stellar evolution calculations of massive stars cannot predict the B[e] supergiant phase (see e.g. Maeder & Meynet [8]; Maeder et al. [9]).

UNCLASSIFIED B[E] STARS: THE EXAMPLE OF HEN 2-90

The enigmatic galactic star Hen 2-90 belongs to the group of unclassified B[e] stars. It has been classified in the literature either as a compact planetary nebula (e.g. Costa et al. [10]; Lamers et al. [1]) or as a symbiotic object (Sahai & Nyman [11]; Guerrero et al. [12]). On an HST image (Sahai et al. [13]) a non-spherical wind structure is visible: a high-ionized polar wind, a low-ionized wind at intermediate latitudes and a (neutral) disk structure in equatorial direction.

We took high- and low-resolution optical spectra with slit centered on this non-spherical wind structure. The spectra contain very strong Balmer lines and lots of forbidden emission lines from e.g. S, N, O, Ar, Cl, Fe in different ionization states. We detected neither emission from HeII nor TiO absorption bands, which are both the main characteristics of a symbiotic object. In addition, we performed a detailed analysis of almost all observed forbidden emission lines and found that C and O are depleted, which means that Hen 2-90 should be an evolved object. Our conlusions are therefore, that Hen 2-90 is most probable a compact planetary nebula (Kraus et al. [14]).

There are, however, a few puzzling details that have to date not been solved: (i) Hen 2-90 shows a jet like structure with many knots being ejected regularly on both sides of the star and perpendicular to the disk. (ii) The velocity profiles of the different emission lines hint to a much more complex structure of the circumstellar material than might be explained with a simple wind model. (iii) We also found from our modeling that N is depleted, which cannot be understood in terms of a normal single star evolution so that probably a binary nature of this star needs to be taken into account (Kraus et al. [15]). More observations are certainly needed to disentangle the nature of this fascinating object.

HOW CAN WE DETERMINE EVOLUTIONARY PHASES ?

The determination of the evolutionary phases of B[e] supergiants and unclassified B[e] stars strongly depends on the interplay of theory with observations. Here, we want to mention some of the methods we are using to fix stellar and circumstellar parameters necessary for the classification.

- **Abundance determinations:**
  The determination of elemental abundances is one key project when dealing with stellar evolution. It is thereby necessary to determine abundances in the circumstellar matter as well as on the stellar surface. The circumstellar material mirrors the surface abundance at the time of matter ejection, while the surface itself gives the actual abundance of the star. The surface abundance can be quite different from the circumstellar abundance if (i) the star was ejecting its complete hydrogen rich outer layers leaving a He rich surface behind, or (ii) the star is rapidly rotating leading to rotationally induced mixing of the internal material which results in a continuous surface enrichement of processed material. Since rapidly rotating stars have high mass loss rates, the mixing leads to an abundance gradient in the wind material.
- **Determination of the geometry and kinematics of the circumstellar matter:**
  The geometry of the system is a crucial point, especially if no direct imaging is available or possible. As discussed in the previous section, the appearance and strength of specific (e.g. forbidden) emission lines can hint to a non-spherical density distributions and even to neutral material close to the hot star. A detailed analysis of forbidden (and permitted) emission lines is needed to derive the ionization structure in non-spherical winds and disks. Especially in the case of B[e] stars we know that in polar direction highly-ionized, line-driven winds of rather high velocity emanate, while from equatorial directions the narrow emission lines of low-ionized or even neutral metals are observed. Both types of profiles will be present in high-resolution spectra.
  Besides the atomic and ionic lines also molecular emission, like e.g. the CO bands can be used to determine the kinematics of the emitting gas. Especially for several B[e] supergiants the CO bands which arise in the near-infrared have been observed. High-resolution observations of the CO $2 \longrightarrow 0$ band head display the complete velocity information of the CO emitting gas. This band head shows e.g. a red peak and a blue shoulder in case of rotation. By modeling the high-resolution, high signal-to-noise $2 \longrightarrow 0$ band head structure one can discriminate contributions coming from rotation and/or outflow (see e.g. Kraus et al. [16]).
- **Determination of the mass loss history:**
  The analysis of the forbidden emission lines can also be used to derive (non-spherical) mass fluxes and therefore the total mass loss rate of a star as has been shown by Kraus et al. [14] for the unclassified B[e] star Hen 2-90. Mass loss rates of the B[e] supergiants and unclassified B[e] stars can then be compared with those available in the literature for stars of different initial conditions and in different evolutionary phases to find agreements and therefore a possible evolutionary stage of each star.

OBSERVATIONS NEEDED

For the modeling of the CO bands high-resolution NIR spectra are needed. PHOENIX, the new high-resolution near-infrared spectrograph at Gemini-South is most suitable for these observations.

For the determination of terminal velocities, the mass loss history and the surface and circumstellar matter abundances we need mainly high-resolution optical (and UV) observations. Since many of the unclassified B[e] stars are southern objects and most of the nowadays known B[e] supergiants are located in the Magellanic Clouds, the new South African Large Telescope (SALT) will be the ideal tool to guarantee the success of our projects. Its major targets will be the Magellanic Clouds (see contributions by David and William in this volume) allowing us to retrieve the high-resolution optical spectra which are needed for a proper classification of B[e] stars.

ACKNOWLEDGMENTS

This research was supported by the Nederlandse Organisatie voor Wetenschappelijk Onderzoek grant No. 614.000.310.

REFERENCES

1. H. J. G. L. M. Lamers, F.-J. Zickgraf, D. de Winter, L. Houziaux and J. Zorec, *Astron. Astroph.* **340**, 117-128 (1998).
2. F.-J. Zickgraf, B. Wolf, O. Stahl, C. Leitherer and G. Klare, *Astron. Astroph.* **143**, 421-430 (1985).
3. M. Kraus and H. J. G. L. M. Lamers, *Astron. Astroph.* **405**, 165-174 (2003).
4. M. Kraus and M. Borges Fernandes, in *The Nature and Evolution of Disks Around Hot Stars*, edited by R. Ignace and K. Gayley, ASP Conference Series, San Francisco: Astronomical Society of the Pacific, 2004, astro-ph/0408073.
5. P. J. McGregor, D. J. Hillier and A. R. Hyland, *Astroph. J.* **334**, 639-656 (1988a).
6. P. J. McGregor, A. R. Hyland and D. J. Hillier, *Astroph. J.* **324**, 1071-1098 (1988b).
7. P. J. McGregor, A. R. Hyland and M. T. McGinn, *Astron. Astroph.* **223**, 237-240 (1989).
8. A. Maeder and G. Meynet, *Ann. Rev. Astron. Astroph.* **38**, 143-190 (2000).
9. A. Maeder, G. Meynet and R. Hirschi, in: *The Fate of the Most Massive Stars*, edited by R. Humphreys and K. Stanek, ASP Conference Series, San Francisco: Astronomical Society of the Pacific, 2004, astro-ph/0408320.
10. R. D. D. Costa, J. A. de Freitas Pacheco and W. J. Maciel, *Astron. Astroph.* **276** 184-186 (1993).
11. R. Sahai and L.-Å Nyman, *Astroph. J.* **537**, L 145-149 (2000).
12. M. A. Guerrero, L. F. Miranda, Y. H. Chu, M. Rodriguez and R. M. Williams, *Astroph. J.* **563**, 883-888 (2001).
13. R. Sahai, S. Brillant, M. Livio, E. K. Grebel, W. Brandner, S. Tingay and L.-Å Nyman, *Astroph. J.* **573**, L 123-127 (2002).
14. M. Kraus, M. Borges Fernandes, F. X. de Araújo and H. J. G. L. M. Lamers, *Astron. Astroph.* (2004) *submitted.*
15. M. Kraus, M. Borges Fernandes and F. X. de Araújo, in: *Massive Stars in Interacting Binaries*, edited by A. Moffat and N. St-Louis, ASP Conference Series, San Francisco: Astronomical Society of the Pacific, 2004, astro-ph/0410196
16. M. Kraus, E. Krügel, C. Thum and T. R. Geballe, *Astron. Astroph.* **362**, 158-168 (2000).

Eclipsing binaries in globular clusters as age and distance indicators

J. Kaluzny*, I.B. Thompson†, W. Krzeminski†, G.W. Preston†, W. Pych*, S.M. Rucinski**, A. Schwarzenberg-Czerny*, S.A. Shectman† and G. Stachowski*

*Nicolaus Copernicus Astronomical Center, Warsaw
†Observatories of the Carnegie Institution of Washington, Pasadena
**University of Toronto, Toronto

Abstract.
We show that DEBs can serve as excellent age and distance indicators for globular clusters. A short review of past surveys for eclipsing binaries in globular clusters is followed by presentation of the current status of the CASE project. A few selected systems from the CASE sample are discussed in some detail.

INTRODUCTION

The ages of Galactic globular clusters (GCs) are of great importance because these are the oldest known systems whose ages can be determined accurately. Measured cluster ages thus place limits on Galactic formation theories, and are a fundamental data point for cosmological studies. Based on first-year observations of WMAP, Spergel et al. (2003) concluded that the age of the universe is 13.7 ± 0.2 Gyr. It is of considerable interest to double-check that result by determining the ages of the oldest stars in the Universe.

The most common method used in deriving ages of GCs exploits the theoretical relation between age and absolute luminosity of stars at the main-sequence turn-off. A condensed discussion of model uncertainties associated with this method can be found in Gratton et al. (2003). Existing measurements of ages of GCs remain controversial mostly because their distances are uncertain. Distances based on the main-sequence fitting method are affected by the lack of a good empirical calibration of M_V versus color for low metallicity stars. Attempts to establish such a relation based on the sample of local unevolved subdwarfs observed by Hipparcos are undermined by the fact that Hipparcos parallaxes suffer from some systematic errors at least for relatively more distant objects (Narayanan & Gould 1999; see also Zwahlen et al. 2004 and references therein). Other potentially accurate distance indicators are red horizontal branch stars, also known as "red clump stars". However, the empirical calibration of their absolute luminosity also suffers from some systematic uncertainties (Salaris, Percival & Girardi 2003).

In this contribution, we discuss how detached eclipsing binaries (DEBs) can be used to derive ages and distances to nearby GCs. A brief discussion of the methods involved is

CP752, *Stellar Astrophysics with the World's Largest Telescopes: First International Workshop*
edited by J. Mikołajewska and A. Olech

followed by summary of the results of some past surveys for DEBs in GCs. We conclude by presenting the current status of the Cluster AgeS Experiment project which aims at detection and detailed analysis of DEBs in nearby GCs.

WHY ECLIPSING BINARIES?

Binary stars are the prime source of accurate data about stellar radii, masses and absolute luminosities. The specific kind of binaries which are most interesting in the context of determining ages and distances to GCs are detached, eclipsing, double-line spectroscopic binaries (DEBs). Their masses and radii can be measured directly with an accuracy of the order of 1-2% for a wide range of spectral types (Andersen 1991). To attain such accuracy for a given system one needs: a) precise light curves with good coverage of both eclipses; b) radial velocity curves allowing determination of $K_1 \& K_2$ semi-amplitudes with an accuracy better than 1%. From the light curve solution one may determine such parameters as inclination i, relative radii $r_{1,2} = R_{1,2}/A$ and relative luminosities $l_{1,2}$. It is advantageous for the light curve solution if the eclipses are relatively deep and it is best if one of them is total. From radial velocity curves one may then obtain (knowing inclination from photometry) masses and the separation A in absolute units. Knowing A we may in turn derive the absolute radii $R_{1,2}$. As discussed below, determination of the age of a given binary requires knowing the absolute bolometric luminosities of its components. Determination of $L_{\rm bol}$ is less direct than determination of masses and radii. One possibility is to estimate an effective temperature based on the dereddened color index or from the spectrum and then use the Stefan-Boltzmann law (the radii are already known). Another way is to first derive the distance modulus $(m-M)_{\rm V}$ for the binary and then apply bolometric corrections to the absolute visual magnitudes M_V of the components. Bolometric corrections are clearly a source of additional uncertainty as they lack accurate empirical calibration for metal poor stars.

Distances

The most direct and least model dependent way to determine distances to DEBs relies on a surface brightnesses method[1]. By definition, the angular diameter of a given star Φ is related to its surface brightness $S_{\rm V}$ and an apparent magnitude V by the relation (Di Benedetto 1998):

$$\Phi[mas] = 10^{0.2(S_{\rm V}-V+A_{\rm V})} \tag{1}$$

where Av is the extinction. A distance and an apparent distance modulus can be derived using the relations:

$$d[pc] = 1.337 \times 10^{-5} R[km]/\Phi[mas] \tag{2}$$

$$(m-M)_{\rm V} = 5 \times log(d) - 5 + A_{\rm V} \tag{3}$$

[1] This method, in a version exploiting the $S_{\rm V}(V-R)$ relation, was originally introduced by Barnes & Evans (1976).

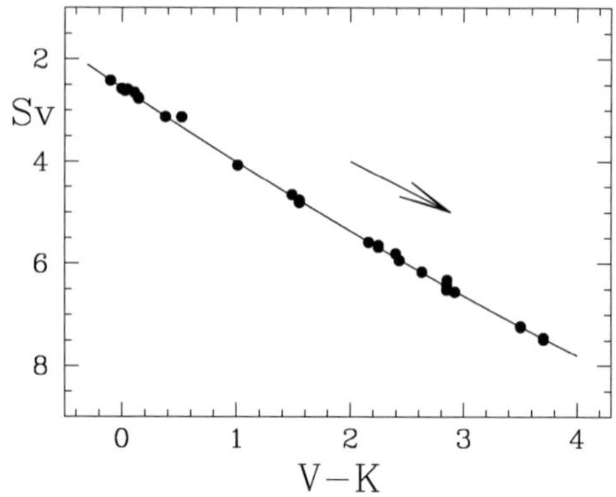

FIGURE 1. Surface brightness of nearby stars in the V band versus $V-K$ color. The arrow corresponds to a reddening $A_V = 1$ mag.

It is worth noting that for a given binary, the distance can be determined separately for each of the two components, which serves as a kind of self-consistency test. Figure 1 shows the $S_V(V-K)$ relation based on the data from Di Benedetto (1998)[2]. The rms value of the adopted fit is 0.023 mag. The reddening vector is essentially parallel to the relation which implies that modest errors in the adopted reddening do not impact the values of derived physical distance (see Eq. 1). By necessity all samples of stars used to calibrate S_V are limited to objects from the solar neighborhood. Fortunately, models of stellar atmospheres indicate that at least the $S_V(V-K)$ relation shows very weak dependence on metallicity (Houdashelt et al. 2000). On the other hand, one has to keep in mind that errors in adopted reddening do affect estimates of the apparent distance modulus.

Ages

For DEBs with components close to the main sequence turn-off point, the age determination is straightforward using the turnoff-mass versus age relation. Note that while the radius of a stellar model is affected by the mixing length parameter for sub-photospheric convection, the luminosity is independent of this uncertainty (Paczyński 1984). The rate of stellar evolution is related to the luminosity, not radius, hence the age determination based on the mass at the turn-off point is very robust. In particular, this measurement of the age is independent of the cluster distance. This method can be illustrated using the data obtained by our group (Kaluzny et al. in preparation) for OGLEGC-17. Figure 2 shows the location of the components of this binary on the color-magnitude diagram for ω Cen. The more massive component of OGLEGC-17 has already passed

[2] Recent empirical calibration of Sv based on VLTI/VINCI data was published by Kervella et al. (2004). These authors use a different definition of surface brightness than that used in Di Benedetto (1998).

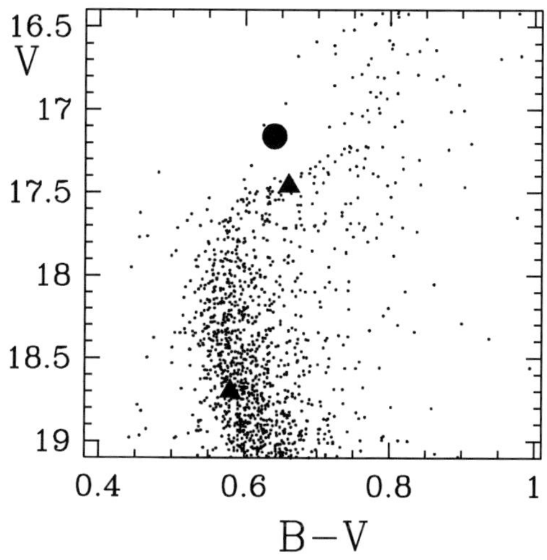

FIGURE 2. The color-magnitude diagram of ω Cen. The location of both components of OGLEGC-17 are marked along with the position of the composite image (large circle).

the turn-off point. From analysis of spectroscopic and photometric data we derived masses and luminosities of both components of the binary: $m_1 = 0.809 \pm 0.012 m_\odot$, $m_2 = 0.757 \pm 0.011 m_\odot$, $L_1 = 3.12 \pm 0.39 L_\odot$ and $L_2 = 0.90 \pm 0.12 L_\odot$. Figure 3 shows a theoretical age/"turn-off mass" relation based on models from Girardi et al. (2000). Assuming that the primary component of OGLEGC-17 is exactly at the turn-off point, we may infer that its age is $\tau = 11.8 \pm 0.6$ Gyr. As can be seen in Fig. 3, the star has already passed the turn-off point and hence its age is a bit higher than the quoted number. Figure 4 shows lines of constant mass on an age/luminosity diagram for $m_1 \pm 1\sigma$ and $m_2 \pm 1\sigma$. Ranges of luminosities $L_1 \pm 1\sigma$ and $L_2 \pm 1\sigma$ are marked with vertical lines. From this plot, one may infer that the age of the binary is about 12 ± 0.9 Gyr. It is also evident from Fig. 3 that in order to get any interesting constraints on the age of DEBs in GCs, one has to measure the masses of their components with a precision of 1% or better. Chaboyer & Krauss (2002) used our results to derive an age of $11.10 + / - 0.67$ Gyr for OGLEGC-17 (see also Chaboyer's contribution in these proceedings).

PAST SURVEYS FOR ECLIPSING BINARIES IN GC

The first ever eclipsing binary belonging to globular cluster was discovered in ω Cen by Niss et al. (1978). It is remarkable that the system, known as NJL5=V192, was detected as one of 83 candidate variables, mostly spurious, identified by inspection of 11 photographic plates taken during commissioning of the 3.6-m ESO telescope. Very promptly Liller (1978) established the period of NJL5 at 1.376 d and published a photo-

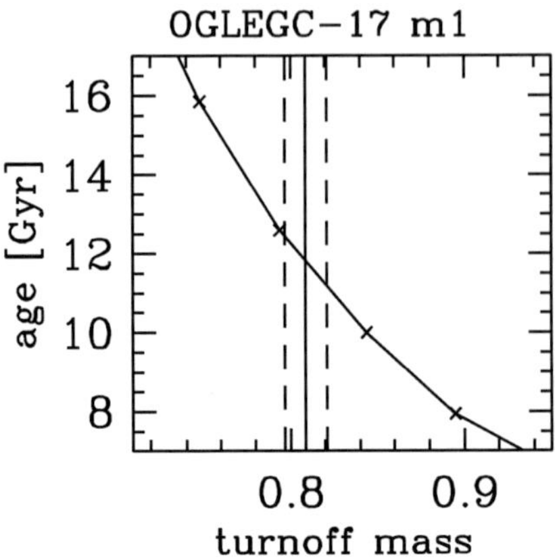

FIGURE 3. Age versus "turnoff mass" relation. Vertical lines represent $\pm 1\sigma$ range for the mass of the primary of OGLEGC-17.

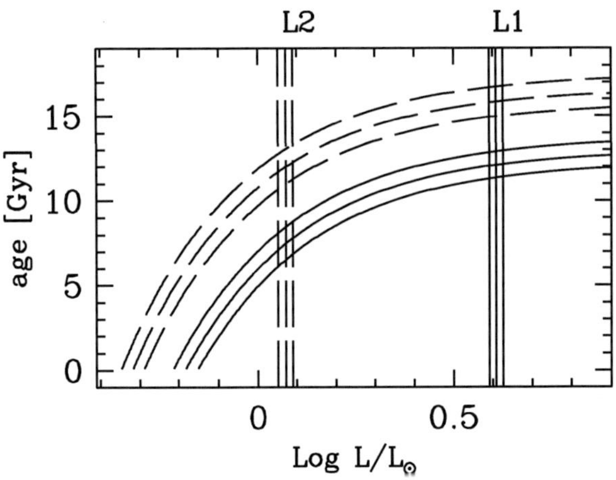

FIGURE 4. Lines of constant mass on age-luminosity diagram. Continuous and dashed lines mark $\pm 1\sigma$ range for the primary and secondary of OGLEGC-17, respectively.

graphic light curve based on 250 plates from the Harvard Observatory collection. Later on, Jensen & Jorgensen (1985) published *BV* light curves based on CCD data. These light curves were analyzed in some detail by Helt et al. (1993). NJL5 is located in the blue straggler area on the color-magnitude diagram of ω Cen and the shape of its light curve suggests a semi-detached configuration of the binary. Therefore, some authors considered it an unlikely member of the cluster. The situation was clarified by Margon & Cannon (1989) who demonstrated that the binary is in fact a radial velocity member

of ω Cen and that it is a single line spectroscopic binary. Up to now, the system awaits a detailed spectroscopic study aimed at determination of its absolute elements.
In the middle of the 1980s, CCD cameras became available at several major observatories. Their advent, connected with development of software suitable for photometry in crowded fields (eg. Daophot (Stetson 1987)) offered the promise of significant progress in the field of variability surveys. However, despite the efforts of some observers, several years passed before the first eclipsing binaries in GCs were detected with the CCD technique. Mateo et al. (1990) reported the identification of 3 EB-type eclipsing blue stragglers in NGC 5466. Later on, Kaluzny & Krzeminski (1993) detected 8 contact binaries and 1 EA-type system in NGC 4372. Yan & Mateo (1994) found 5 EW/EB binaries in M71. All binaries from the samples mentioned above are either blue stragglers or have periods $P < 1$ d. Hence, they are not useful as age indicators. Rich samples of eclipsing binaries were detected in ω Cen and 47 Tuc by the OGLE team (Kaluzny et al. 1996; 1997; 1998). The OGLE sample included in particular the first known main-sequence DEBs in GCs. As discussed above one of these binaries – OGLEGC-17 – has been the subject of detailed follow up studies. Recently Weldrake et al. (2004) reported detection of a few new DEBs located in the outer parts of 47 Tuc.

Ground based surveys for DEBs in GCs are demanding mostly in respect of extensive time coverage. Also reasonable seeing is desirable and it becomes a crucial factor if one aims at detection of variables in the central crowded regions of GCs. It is known that due to mass segregation effect binaries tend to be located in the cores of GCs. One may overcome the limitations imposed by seeing and crowding effects by using the Hubble Space Telescope. An excellent example of what can be accomplished with the HST in that respect is a study of the central part of 47 Tuc by Albrow et a. (2001). These authors analyzed the WFPC2 images with near-continuous coverage over a time base spanning 8.3 days. As a result, they detected 11 DEBs and 15 contact binaries. The data were in fact collected with the primary goal of looking for planetary transients in the cluster. It would be rather difficult to obtain an allocation of interestingly large number of orbits on the HST to survey other GCs for DEBs. Some attempts were made to use archival HST/WFPC2 data to search for eclipsing binaries in GCs but the detected sample is limited mostly to contact binaries (Pietrukowicz & Kaluzny 2004). Moreover one has to keep in mind that binaries located in cores of GCs are difficult targets for ground-based spectroscopic follow-up observations. The HST is not helpful in that respect because of its limited spectroscopic capabilities.

THE CASE PROJECT

In 1996 we started a long term project named "Cluster AgeS Experiment". It aims at determination of ages and distances to nearby GCs by using observations of DEBs. The survey part of the project has been conducted using the 1.0-m Swope telescope at Las Campanas. We were allocated over 250 nights during the period 1996-2003 and used them to monitor about a dozen GCs. Follow-up optical and IR photometry of detected binaries has been obtained with the 2.5-m DuPont telescope. Precise multi-color photometry in particular allowed us to drop some objects from the sample as they turned out to be field stars. Whenever possible, we use archival HST images to check

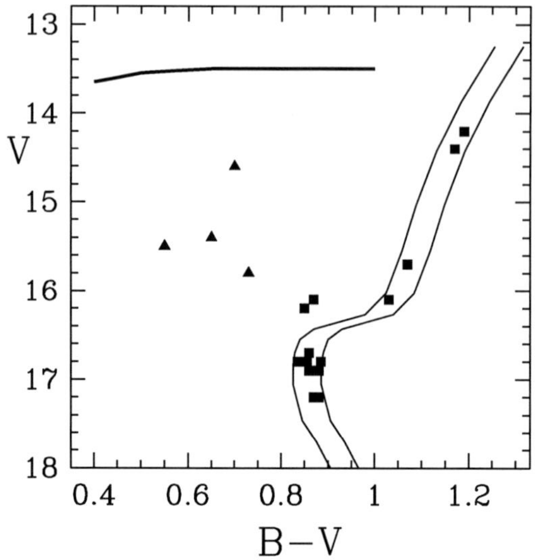

FIGURE 5. CASE sample of double-lined eclipsing binaries that are confirmed members of globular clusters. Squares denote detached turnoff and subgiant branch binaries. Triangles are detached or semi-detached blue stragglers.

against the presence of possible close optical companions to binaries from the sample. In 2002 we observed several binaries from our list with the medium resolution B&C spectrograph attached to the 6.5-m Baade telescope. These observations were aimed at determination of systemic velocities for objects located in clusters with high radial velocities. It was possible to further clean the sample from field objects. Since 2003 we are in the process of determining radial velocity curves for several DEBs with help of the high resolution MIKE spectrograph mounted permanently on the 6.5-m Clay telescope. In most cases a few data points are sufficient for determination of systemic velocity for a given binary with accuracy allowing determination of its membership status. Proven clusters members are observed further with the goal of obtaining the masses of their components with accuracy better than 1%. Below we present some results obtained so far by the CASE collaboration.

Figure 5 shows newly discovered double-lined DEBs that are confirmed members of GCs. The variables are plotted on a schematic color-magnitude diagram for the globular cluster M4. Appropriate difference in distances and reddenings were taken into account. Our current sample includes 4 blue stragglers which can be used as distance indicators. The remaining 14 binaries are composed of main–sequence stars or subgiants. Their membership status is based on the systemic velocity criterion as well as on the location of their components on the color-magnitude diagrams of parent clusters.

Our sample includes three DEBs belonging to the closest globular cluster, M4. Variable M4-V11888 has an orbital period of about 8 days. It is composed of two upper main-sequence stars what can be seen in Fig. 6. The light curve of V11888 is presented

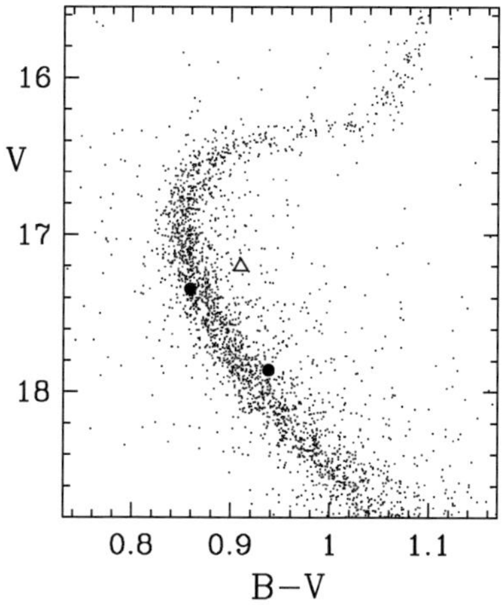

FIGURE 6. Color-magnitude diagram of M4. Large circles indicate components of M4-V11888. Triangle indicates position of M4-V23900.

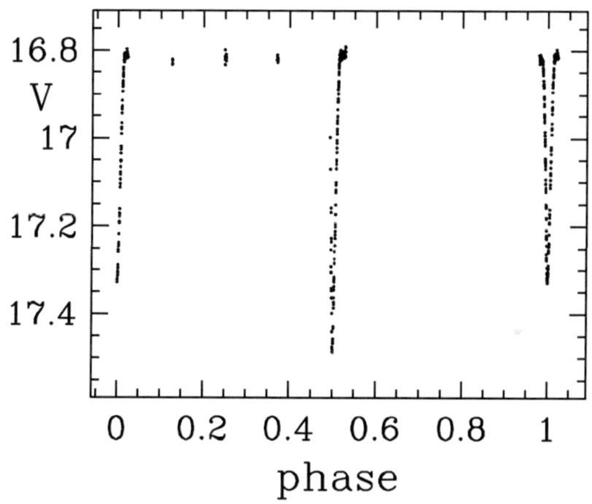

FIGURE 7. Light curve of M4-V11888

in Fig. 7. Its shape implies that the orbital inclination of the system is very close to 90 deg. The binary has a circular orbit and its shows two eclipses of comparable depth. The systemic velocity equals +78 km/s which can be compared with $V_{rad} = 70.4 \pm 3.5$ km/s observed for M4 stars.

The second binary marked in Fig. 6 has a period around 46 days and was found from spectroscopic observations. The radial velocity curve of M4-V23900 with a preliminary orbital solution is shown in Fig. 8. So far we have observed only one eclipse event for

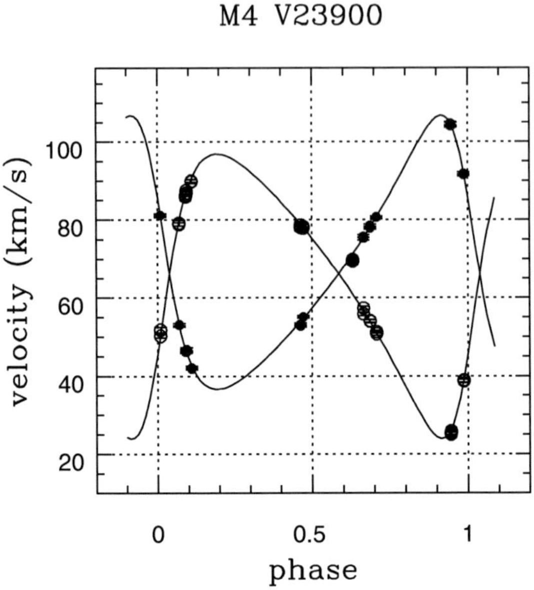

FIGURE 8. Radial velocity curves of components of M4-V23900.

this binary but it should be possible to establish an accurate ephemeris once more spectra are in hand. That in turn will allow planning of photometric observations aimed at covering the light curve in eclipses.

We obtained dedicated follow-up observation of an eclipsing blue straggler OGLEGC-14=V209 located in ω Cen. This variable attracted our attention because its color is very blue, even for a blue straggler, $(B-V)_0 = 0.06$. Moreover, the light curve is rather flat out of eclipse, which can be seen on Fig. 9. Closer examination reveals that the primary eclipse is total, which helps in obtaining a unique and accurate light curve solution. Figure 10 shows the location of components of V209 on the color-magnitude diagram of the cluster. The hotter and less luminous component occupies a position corresponding to sdB/O subdwarfs, while the cooler and brighter is located somewhere between the EHB branch and hot subdwarfs. The variable is a certain radial velocity member of ω Cen. [3] It seems that V209 is a close but detached binary composed of two hot subdwarfs whose absolute parameters can be derived with high accuracy once we finish collecting the necessary spectroscopic data.

We have detected only one detached eclipsing binary in NGC 362. The variable NGC362-V11 is very interesting as it is likely composed of two subgiant stars. Its location on the cluster color-magnitude diagram is shown in Fig. 11. The variable is present on some archival HST/WFPC2 images and hence we are confident that our

[3] The cluster has $V_{\rm rad} = +232$ km/s which makes identification of field interlopers straightforward.

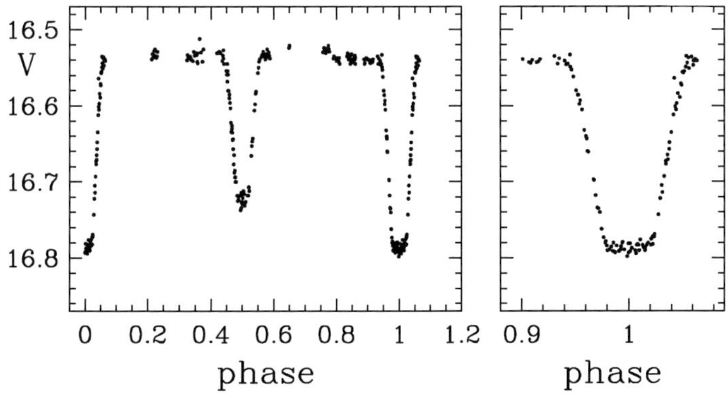

FIGURE 9. Light curve of binary subdwarf V209.

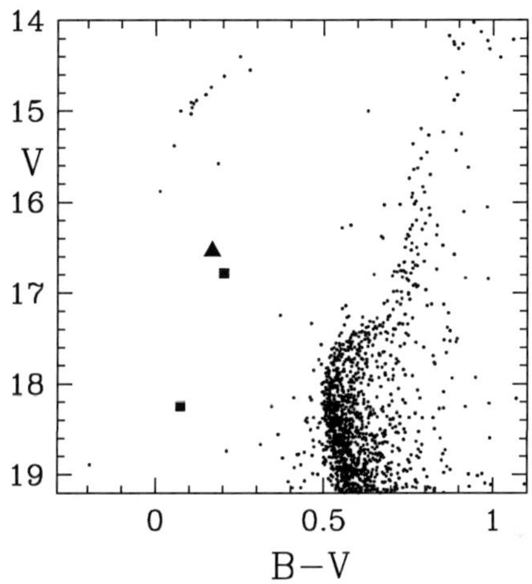

FIGURE 10. Color-magnitude diagram of ω Cen. The locations of both components of V209 are marked along with the position of the composite image (triangle).

ground-based photometry is free from any blending effects. Its orbital period is about 8 days and the light curve from the 1999 season is presented in Fig. 12. One may note the disturbed shape of the light curve which is most likely related to chromospheric activity. Considering its spectral type the binary can be classified as a RS CVn type system. The presence of eclipses indicates that the orbital inclination is close to 90 deg. However, detailed analysis of the system is complicated because of the disturbed shape of its light curve. The preliminary radial velocity curves of NGC362-V11 are shown in Fig. 13. One may see that the mass ratio of the binary is very close to unity. We are hoping that analysis of light curves from different seasons will allow us to tightly constrain the orbital inclination of the system, leading to a determination of accurate masses of its

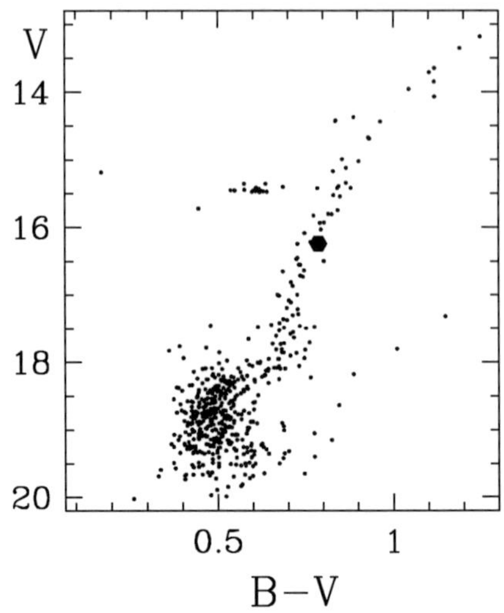

FIGURE 11. Color-magnitude diagram of NGC 362. The location of V11-N362 is marked with a diamond.

components.

The core of 47 Tuc is known to harbor several exotic objects (Edmonds et al. 2003). It turns out, however, that some interesting objects also exist in the outer parts of the cluster. An eclipsing blue straggler OGLEGC-228 (Kaluzny et al. 1998), is an unblended star located about 5.3 arcmin form the cluster center. It is a highly probable proper motion member of the cluster (Tucholke 1992). Our data give a systemic velocity of the binary as $\gamma = -23.5$ km/s, which can be compared with the cluster velocity equal to -18.7 km/s (Gebhardt et al. 1995; from their data velocity dispersion at $d = 5$ arcmin can be estimated at 8 km/s). The light curve of OGLEGC-228 is shown in Fig. 14. Analysis of the light curve indicates a semidetached or marginally detached configuration of the binary. The radial velocity curves presented in Fig. 15 were extracted using the BF method as described by Rucinski (2002). Our preliminary solution, based on the available photometry and spectroscopy indicates that the total mass of the binary is $m_1 + m_2 > 1.6 \pm 0.2 m_{\odot}$. We hope to obtain more complete coverage of the radial velocity curves of OGLEGC-228 soon and to present an accurate determination of its absolute parameters.

SUMMARY

Detached eclipsing binaries can serve as excellent distance and age indicators for globular clusters. To obtain interesting constraints on ages, one has to be able to measure masses of binary components with an accuracy better than 1%. Such a level of accuracy

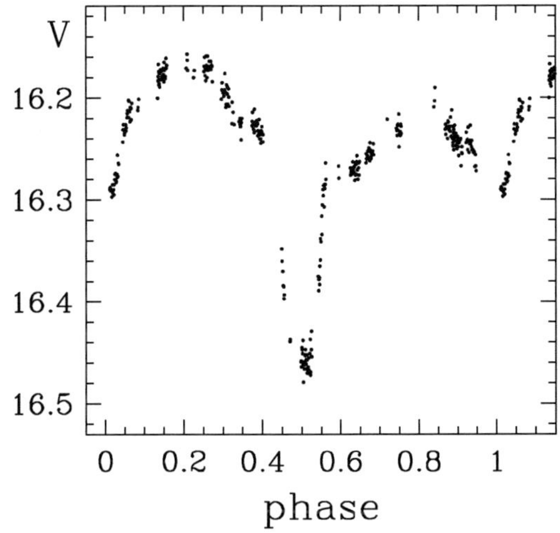

FIGURE 12. Light curve of binary subgiant NGC362-V11.

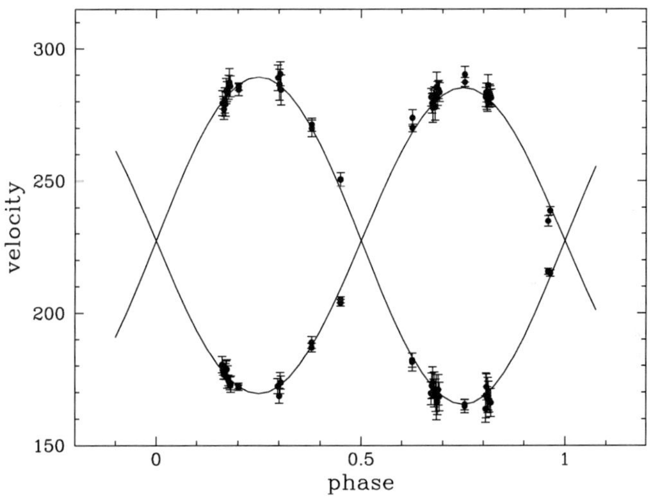

FIGURE 13. Radial velocity curves of components of NGC362-V11.

can be obtained with modern instruments for turn-off binaries in nearby clusters. Near IR photometry of target binaries is necessary if one aims at determination of accurate distances. Several DEBs with periods ranging from about 1 day up to 46 days were detected in globular clusters by the CASE collaboration. We are in the process of deriving radial velocity curves as well as follow-up photometry for these objects.

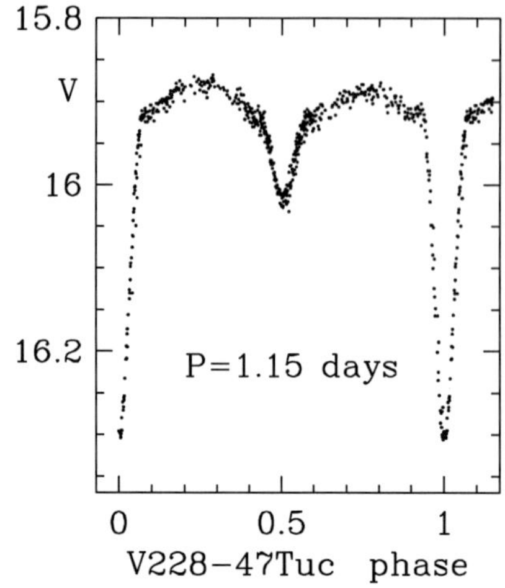

FIGURE 14. Light curve of OGLEGC-228

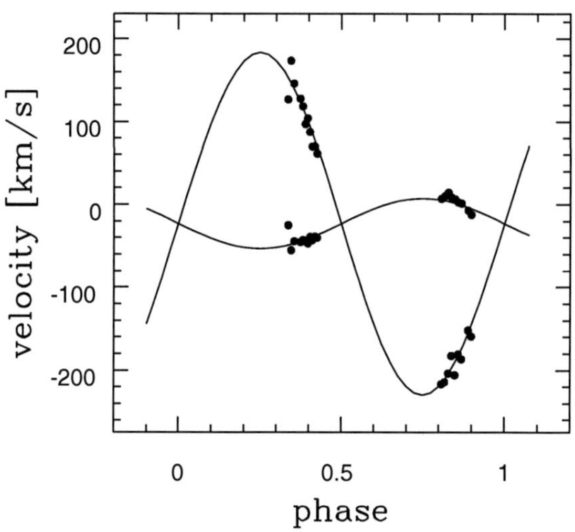

FIGURE 15. Radial velocity curves of components of OGLEGC-228.

ACKNOWLEDGMENTS

The CASE project has been conceived by Bohdan Paczyński. We would like to express our deepest thanks for his support and valuable input at different phases of the project.

REFERENCES

1. Albrow, M., Gilliland, R. L., Brown, T., Edmonds, P. D., & Sarajedini, A. 2001, ApJ, 559, 1060
2. Andersen, J. 1991, A&AAR, 3, 91
3. Barnes, T.G., &Evans, D.S. 1975, MNRAS, 174, 489
4. Chaboyer, B., Krauss, L. M. 2002, ApJ, 567, L45
5. Di Benedetto, G.P., A&A, 339, 858
6. Edmonds, P. D., Gilliland, R. L., Heinke, C. O., Grindlay, J. E. 2003, ApJ 596, 1197
7. Gebhardt, K., Pryor, C., & Williams, T.B. 1995, AJ, 110, 1699
8. Girardi, L., Bressan, A., Bertelli, G., & Chiosi, C. 2000, A&AS, 141, 371
9. Gratton, R.G. et al. 2003, A&A, 408, 529
10. Groenewegen, M. A. T., Hummel, C. A. 2004, A&A 425, L45
11. Helt, B. E., Jorgensen, H.E., King,S., Larsen, A. 1993, A&A, 270, 297
12. Houdashelt, M.L., Bell, R.A., & Sweigart, A.V. 2000, AJ 119, 1448
13. Jensen, K. S., Jorgensen, H.E. 1985, A&AS, 60, 229
14. Kaluzny, J., & Krzemiński, W. 1993, MNRAS, 264, 785
15. Kaluzny, J., Kubiak, M., Szymanski, M., Udalski, A., Krzeminski, W., & Mateo, M.M. 1996, A&AS, 120, 139
16. Kaluzny, J., et al. 1997, A&AS, 122, 471
17. Kaluzny, J., et al. 1998, A&AS, 128, 19
18. Kervella, P., Thévenin, F., Di Folco, E., Ségransan, D. 2004, A&AS, 426, 297
19. Liller, M. H. 1978, Information Bulletin on Variable Stars, Vol. 1527
20. Margon, B., Cannon, R. 1989, Observatory, 109, 82
21. Narayanan, V.K. & Gould, A. 1999, ApJ, 523, 328
22. Niss, B., Jorgensen, H. E., Laustsen, S. 1978, A&AS, 32, 387
23. Paczyński, B. 1984, ApJ, 284, 670
24. Pietrukowicz, P., & Kaluzny, J. 2004, AcA, 54, 19
25. Rucinski, S. M. 2002, AJ, 124, 1746
27. Salaris, M., Percival, S., & Girardi, L. 2003, MNRAS, 345, 1030
27. Spergel, D.N. et al. 2003, ApJS, 148, 175
28. Stetson, P. B. 1987, PASP, 99, 191
29. Tucholke, H.-J. 1992, A&AS, 93, 293
30. Weldrake, D. T. F., Sackett, P. D., Bridges, T. J., Freeman, K. C. 2004, AJ, 128, 736
31. Yan, L., & Mateo, M.M. 1994, AJ, 108, 1810
32. Zwahlen, N., North, P., Debernardi, Y., Eyer, L., Galland, F. 2004, A&A, 425, L45

The LAMOST/Galaxy Project

Jingyao Hu* and Biwei Jiang†

*National Astronomical Observatory, Chinese Academy of Sciences, Beijing 100012
†Department of Astronomy, Beijing Normal University, Beijing 100875

Abstract. LAMOST is one of the Chinese national big science programs. As an all reflecting Schmidt telescope with an effective aperture of 4 meters, LAMOST has a field of view of 5 degrees. Its 4000 optical fibers can point to selected celestial objects automatically and link to 16 spectrographs, thus to obtain about 4000 spectra simultaneously. With the large field of view and multi-fibers, LAMOST is efficient to collect huge number of stellar spectra , which leads us to propose using it to study our Milky Way galaxy.

THE LAMOST PROGRAM

In the beginning of 1990 Wang et al (1996) raised a new concept of all reflecting Schmidt telescope in which both the primary mirror and corrector are reflectors, and sent a proposal to the Sciences and Technologies Group of Chinese State Council to build a such telescope, i.e. LAMOST (The Large Sky Area Multi-Object Fiber Spectroscopic Telescope). In August 2001, this proposal was approved by Chinese State Council and the LAMOST program became one of the national big science programs.

The LAMOST telescope has an aperture of 4m, focal ratio of 5, and a 5deg field of view. Its optical layout is shown in Fig.1. In comparison with previous Schmidt telescopes, LAMOST has some innovations. The basic facts and main innovations are summarized as following.

- LAMOST, an all-reflecting Schmidt telescope, works as a meridian instrument. The spherical primary mirror (Mb in Fig.1) is fixed on the foundation. The reflecting corrector (Ma in Fig.1) is located at the curvature center of Mb. The objects are observed for 1.5 hours as they pass through the meridian. The alt-azimuth mount is adopted for Ma and done the tracking. The optical surface of Ma is changed using active optics technology to correct the spherical error of Mb.
- Ma is a reflecting Schmidt corrector. It consists of 24 hexagonal submirrors. Each segment is 1.1meter long measured across the diagonal and 25mm thick. The main mirror Mb is composed of 37 hexagonal submirrors each of which is also 1.1 meters long across the diagonal. As a whole, the effective aperture of the telescope is about 4 meters. The optical parameters are: f-ratio of 5, focal length of 20 meters, field of view of 5 degree. After the full correction by Ma, about 80% energy will be within 2 arcsec.
- The large focal plane has a diameter of 1.75 meters and is divided into 4000 blocks. Each block is equipped a computer controlled optical fiber pointer. Within 10 min the pointer can move to next object with an accuracy of 0.4 arcsec.

CP752, *Stellar Astrophysics with the World's Largest Telescopes: First International Workshop*
edited by J. Mikołajewska and A. Olech

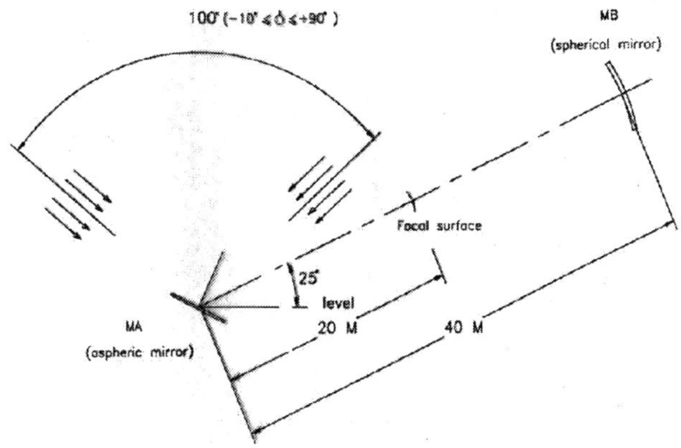

FIGURE 1. Optical system of LAMOST

- 4000 optical fibers are linked to 16 spectrographs. The spectrographs has two working modes. One mode is Low Resolution Spectroscopy mode and the other is Medium Resolution Spectroscopy mode. In the LRS mode, the spectral resolution is about 2000, and the spectral coverage is divided two parts: the blue one is from 380 to 590 nm and the red one from 560 to 900 nm. In the MRS mode, the spectral resolutions are 4000 for the blue band (from 370 to 590 nm) and 10000 for the red band(covering the CaII triplets from 830 to 890 nm). Both modes use an identical spectrograph but different VPH gratings.

THE LAMOST/GALAXY PROJECT

The design of LAMOST is optimum to a very high spectrum acquiring rate as ten-thousands spectra per night. In addition, with the 4-meter effective aperture, LAMOST is available to conduct a high sensitivity all-sky survey. At the Xinglong Observatory of National Astronomical Observatories of China where LAMOST will be put, by integrating for 1.5 hours, the limiting magnitude is 20.5 mag in visual band in dark nights. On the other hand, in moon nights, we can not reach such limiting magnitude (Liu, Zhou et al. 2003). In order to efficiently make use of both the dark and moon nights, we propose to perform two surveys in two modes. One mode is a survey for all the stars with declination northern of -10deg and brighter than V=16 mag with a resolution of R=10000 (i.e. MRS mode). Another mode is a survey for stars in some particular sky regions with a lower resolution (i.e. LRS mode). In the following, the details of the two modes are described.

MRS mode

In this mode, the spectra are separated into two parts: the blue part covers 370-590 nm with a spectral resolution of about 4000, and the red part covers 830-890 nm, i.e. the CaII triplets region. For a 30 min exposure the sensitivity reaches V=16 with S/N>20. The blue part of the spectra will be used to study mainly O,B stars, interstellar matter, and some kinds of peculiar objects. Under this resolution and signal-to-noise ratio, the stellar radial velocity and other atmospheric parameters can be derived with reasonable accuracy from the red part of the spectrum. According to the report of the RAVE pilot survey, stellar radial velocity can be obtained with the precision of approximate 2km/sec. The effective temperature, surface gravity, and metallicities of the A0 to K0-type stars can be determined with an accuracy like $\sigma(T_{\rm eff}) \sim 100$, $\sigma(\log g) \sim 0.4$ dex, and σ([Fe/H])$\sim$0.2. The estimation of [α/H] is feasible at least down to metallicities as low as [Fe/H]= -2.

To consider the time required to complete the survey, lets first have a look at the number of moon nights and the number of stars to be observed. On one hand, there are about 250 clear nights per year at Xinglong Observatory and half of them (i.e. 125 nights) are moon nights. On the other hand, there are about 2.5×10^7 stars brighter than V=16 with declination northern of $-10°$. If we assume that the effective observing time is 8 hours per night on average and the efficiency of optical fiber usage is 90% (i.e. 3600 fibers work well simultaneously), then in one year the system can observe about 7.2×10^6 stars. Therefore it will take 3.5 years to finish the MRS mode all sky survey. As is planned, from 2008 LAMOST will start routine operation, so before 2012 this MRS mode survey will be finished.

The scientific goals of the MRS mode survey is complement and synergistic to that of RAVE. The theme is to study the composition, formation and evolution of our Milky Way galaxy. The subjects include the large scale structure of our Galaxy, such as the degree of ellipticity, warping and the vertical structure of the thin and thick disks; and the small scale structure of our Galaxy such as structure of the nearby spiral arms, local surface density in the thin and thick disks, stellar associations and present star forming regions, structure of the Gould belt. In particularly, we will pay much attention to searching for the stellar streams in the halo and thick disk by using their unique chemical and kinematic signatures left from their accreted parent satellite. From millions halo stars we can expect to find about a thousand stars with [Fe/H]< -4, and ten thousand stars with [Fe/H]< -3, which can be use to characterize the chemical evolution of our Galaxy. In combination with photometric data from such as SLOAN and 2MASS, the LAMOST spectra data can be used to study interstellar matter, its space distribution and interstellar extinctions in various sightlines at a high spatial resolution.

As our survey is non-biased to stars bright than 16 mag in visual bands, it will certainly provide serendipitous discoveries of unanticipated phenomena.

LRS mode

As described above, the MRS mode survey is limited to V=16 and thus can not reach deep in the Galaxy particularly in the direction of dense interstellar regions. In order to attain a deeper view of our Galaxy, the LRS mode is proposed to carry out a survey in some selected regions. In the LRS mode, the spectral resolution is about R=2000, comparatively lower. According to the experience of SLOAN, this resolution should achieve the radial velocity accuracy of about 15 km/s, and $\sigma(T_{\rm eff}) \sim 200$ K, $\sigma(\log g) \sim 0.4$ dex and σ([Fe/H])~0.4 dex. In compensation for the lower resolution, the limiting magnitude for 1.5 hours can reach V=20.5. In this mode the OB stars and red giants can be observed to the edge of the Galaxy even for the regions where A_V=10. These regions are difficult to reach even by the GAIA mission. So the strategy in selection of the survey regions will be decided after referencing the results of SEGUE and GAIA mission.

REFERENCES

1. Wang, S.G., Su, D.Q., Chu, Y.Q., Cui, X.Q., & Wang, Y.N., Applied Optics, 35, 515 (1996)
2. Liu, Y., Zhou, X., et al., PASP 115, 495 (2003)

Star Counts and Galactic Structure: The distribution of main sequence stars and giants in color-color diagram

Xu Yan* and Deng Li Cai*

*National Astronomical Observatories, Chinese Academy of Sciences, Beijing 100012, China

Abstract. Main sequence stars have good color-magnitude relation to obtain distance which is the key to determine the structure of the Galaxy using the method of star counts. A model to mimic observation is made in this paper to infer a certain criteria to distinguish main sequence stars and giants.

INTRODUCTION

Star counts is a classical and the most direct method to study our Galaxy. Numerous released data of SDSS gives us the chance to study the structure of our Galaxy further. Because of good color-magnitude relation to obtain distance, main sequence star can be a tracer to determine structure of the Galaxy directly. So the first step is to distinguish main sequence stars and giants. But this is difficult to realize using observational data only.

By using theory of stellar evolution and a reasonally good model of the Galaxy, we can simulate the result of observation and give a reference to real observation or data reduction. So we created a theoretical sample to simulate distribution of stars in color-color diagram. The goal of this work is try to distinguish theoretically giants and main sequence stars in SDSS photometric system. Some preliminary result has been obtained.

In section 2, the construction of the model and result are presented. In section 3, the result of model is compared with observations. Section 4 draws conclusions of present work.

CONSTRUCT SYNTHETIC MODEL

The model is constructed with Robin et al. (2003)'s model which gives the information of populations in the Galaxy, Bahcall et al. (1980) and Chen et al. (2001)'s star counts model and Girardi et al. (2004), Marigo et al. (2001)'s isochrones in SDSS photometric system.

If stars in spatial angle in some direction are to be simulated, population components should be known a priori. Robin et al. (2003)'s model presented in Table1 gives us the information of age, metallicity, initial mass function, local mass density of different populations. In his model, the thin disk is divided 7 groups and is assumed to form

CP752, *Stellar Astrophysics with the World's Largest Telescopes: First International Workshop*
edited by J. Mikołajewska and A. Olech

TABLE 1. Age, metallicity ([Fe/H])(mean and dispersion about the mean), radial metallicity gradient(dex/kpc), initial mass function (IMF), and star formation rate (SFR) of the stellar components (Robin et al. 2003).

Group	age(Gyr)	[Fe/H]	$\rho_0(M_\odot PC^{-3})$	IMF
1	0-0.15	0.01±0.12	4×10^{-3}	dn/dm=$m^{-1.6}(M<1M_\odot)$
2	0.15-1	0.03±0.12	7.9×10^{-3}	dn/dm=$m^{-3}(M\geq 1M_\odot)$
3	1-2	0.03±0.1	6.2×10^{-3}	
4	2-3	0.01±0.11	4×10^{-3}	
5	3-5	−0.07±0.18	5.8×10^{-3}	
6	5-7	−0.14±0.17	4.9×10^{-3}	
7	7-10	−0.37±0.2	6.6×10^{-3}	
8	11	−0.78±0.3	1.34×10^{-3}	dn/dm=$m^{-0.5}$
9	14	−1.78±0.5	9.32×10^{-6}	

continuously.The thick disk is described by 8th group, while the halo by 9th. They are assumed to form just in a single time.

Number of stars of each population in the spatial angle is calculated using Bahcall et al. (1980) and Chen et al. (2001)'s double exponential disk model.

Equation (1) is the function of star counts. The traditional way is to integrate density profile and luminosity function to obtain number of stars of each magnitude bin. But the number of stars of each mass bin is not yet known. So luminosity function in the equation of Bahcall et al. (1980) is replaced by a mass function. Initial mass function in table 1 is adopted to proximate present mass function. For upper mass limit, the largest mass of stars that are still alive at that age is adopted. In addition, the cone of solid angle is divided to several layers by distances. So that the integration of R is divided into parts. The distance of the center of each layer is regarded as the distance of stars in this layer.

$$A_i = \int_{mass1}^{mass2} dmass \int_R \rho_i(R)\phi_i(mass)d\Omega R^2 dR \qquad i=1,......,9 \tag{1}$$

From star counts model, we can derive star numbers for each population of table 1 in each mass bin, each layer. According to the number, random points are produced in the scope of age and metallicity of Table 1. Age follows even distribution, while metallicity satisfies Gaussian distribution.

At this stage, mass, age and metallicity of each star has been obtained. According to these parameters, effective temperature, luminosity, surface gravity and magnitude in each passband can be interpolated from the isochrones. In this paper, isochrones in SDSS photometric system are used that are given by Marigo et al. (2001)'s metal free isochrone and Girardi et al. (2004)'s isochrones with 7 metallicities from Z=0.03 to 0.0001.

Figure 1 shows the preliminary results. In this diagram, stars of 100 square degrees are simulated whose center is $RA = 180°$, $DEC = 60°$, the integration goes from 0 to 15 kpc. Gray dots describe main sequence stars, black points represent scope of giants. As for these theoretically produced stars, an aperture magnitude limit from 15 to 22

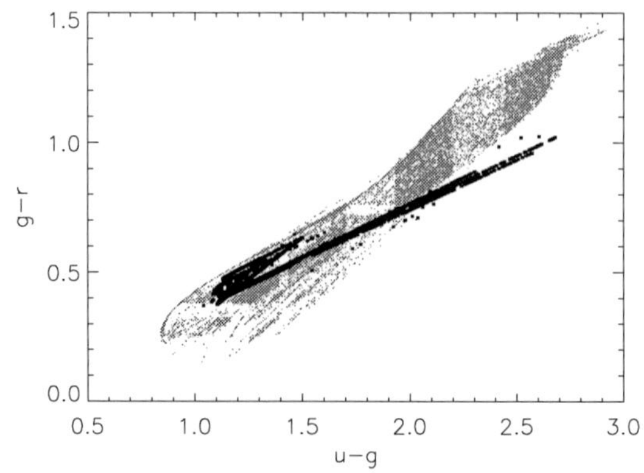

FIGURE 1. Color-color diagram of theoretical model

mag is used based on SDSS observational magnitude limit. Stars outside this limit are filtered out. In this diagram, most of giants and main sequence stars are overlapped. We can only give a scope of giants concentration. The strips of high concentration are from (1.05,0.41) to (1.48,0.62) and from (1.05,0.37) to (2.6,1.0).

In the top left panel of Figure 1 of Helmi et al. (2003), nine Spaghetti Project Survey giants given by Dohm-Palmer et al. (2001) locate around (1.5,0.65). All of these giants are metal-poor halo giants. Most of them seem redder than giants produced theoretically as shown in Figure 1 of this paper.

COMPARE THE RESULT WITH SDSS DATA

Now, we compare the distribution of stars in color-color magnitude between theoretical result and data of SDSS. Theoretical data and observational data has same sky coverage that is 9 square degrees centered in $RA = 180°$, $DEC = 60°$. Left panel of Figure 2 is color magnitude diagram of these observational data, stars beyond 22 magnitude or lower than 15 magnitude have large scatter because of large magnitude error or effect of saturation. So we chose only stars between the limits. The theoretical stars are adopted in the same standard. The theoretical result and observational data are compared in right panel of Figure 2. In this color-color diagram, contour describe theoretical result. The levels are number of stars of 10, 50, 100, 200 in grid of 0.1×0.1. The dots describe SDSS observational data shown in left panel of Figure 2 in the SDSS magnitude limit. The theoretical result and observational result can fit generally. In the right upper corner, the contour can't cover the observational points. Because the low mass limit of isochrone is 0.15 solar mass or 0.2 solar mass, stellar mass lower than this limit can't be included in our model. In the left lower corner, a lot of stars leak from contour. Because they are bluest stars among observational data, most of them are white dwarfs which also cannot be included in the current model.

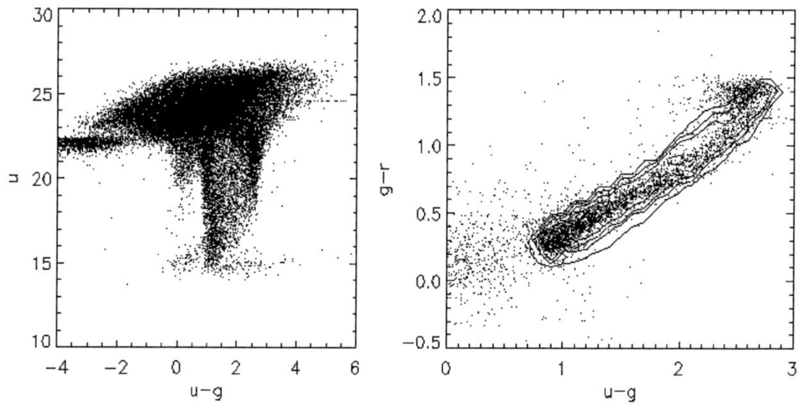

FIGURE 2. The left panel: color magnitude diagram of the observational points. The right panel: In this color-color diagram, contour describe theoretical result, dots describe observational points in the left panel and satisfying SDSS observational limit.

CONCLUSION

The present work can be concluded in the following. Our model generally fit observation in color-color diagram. Giants and main sequence stars cannot be separated photometrically as inferred by our theoretical model, only strips in the color-color diagram with high giant concentration can be given. The observational metal poor giants given by Helmi (2003) are redder than the scope of gaints obtained theoreticlly. Further spectroscopic observation are needed to constrain the theoretical model.Theoretical model with more reliable physics and observations is undertaken.

REFERENCES

1. Bahcall, J. N., Soneira, R. M, ApJS, 44, 73 (1980)
2. Chen, Bing, et al., ApJ, 553, 184 (2001)
3. Girardi, L. et al., astro-ph/0404358
4. Helmi, A. et al. , ApJ, 586, 195 (2003)
5. Marigo, P., Girardi, L.,A&A,371,152 (2001)
6. Robin, A. C. et al., A&A, 409, 523 (2003)

END PRODUCTS OF SINGLE AND BINARY STELLAR EVOLUTION

Large Telescopes and Asymptotic Giant Branch Stars

Patricia Whitelock[1] and Michael Feast[2]

1. South African Astronomical Observatory, PO Box 9, 7935 Observatory, South Africa
2. University of Cape Town, Rondebosch 7700, South Africa

Abstract. There are many ways in which observations with large telescopes are improving our understanding of evolution on the Asymptotic Giant Branch. Perhaps more importantly AGB stars also offer huge potential to probe the nature of low- and intermediate-mass populations over a very considerable range of distances and in a variety of environments. This review which is far from exhaustive, touches upon: the discovery of AGB stars in resolved stellar populations and their detailed investigation; the importance of abundances measurements – particularly those of technetium and lithium; the potential of kinematic studies and recent developments in interferometry. Finally the important role of AGB stars as probes of distant and unresolved populations is mentioned.

THE IMPORT OF AGB STARS FOR ASTROPHYSICS

Low- and intermediate-mass stars (about 1 to $8M_{\odot}$) are, and have been, a major component of the Universe. They die as a result of losing most of their nuclear fuel in a "superwind" as they approach the top of the asymptotic giant branch, leaving behind a white dwarf remnant with a mass in the range 0.6 to 1.4 $M_{\odot}$. The mass loss process remains somewhat enigmatic despite intensive observational and theoretical efforts over the last 10 to 20 yrs. As pointed out by Wood et al. (Spitzer GO-1 program 3505) mass loss from AGB stars is responsible for roughly half the gas recycled by stars, and AGB stars are major contributors to the chemical enrichment, or pollution, of the universe. In particular they are the only confirmed source of primary nitrogen, and produce roughly half the carbon – the rest coming from supernovae and WR stars. They are also the main source of heavy elements made by slow neutron capture and the only known stellar source of lithium. Low- and intermediate-mass stars are at their most luminous during this rather brief evolutionary phase, and in relatively nearby galaxies it is the AGB stars that can be resolved and studied as individual objects. In more distant systems the AGB population makes major contributions to the surface brightness fluctuations which are a potentially important tool not only for studying distance, but also for insight on the age and chemical composition of unresolved populations (e.g. Cantiello et al. 2003). A detailed review of our understanding of AGB stars can be found in the book edited by Habing & Olofsson (2003).

Of special interest are the stars at the very tip of the AGB, sometimes known as Mira variables. The properties of these stars which render them particularly interesting in the context of large telescopes are as follows:

CP752, *Stellar Astrophysics with the World's Largest Telescopes: First International Workshop*
edited by J. Mikołajewska and A. Olech

1. They are very luminous, bolometrically (L >2.9 10^3 $L_\odot$) and at near-infrared (*JHKL*) wavelengths, so they stand out in any old or intermediate age population.
2. Their amplitudes are large ($\Delta V \geq 2.5$, $\Delta K \geq 0.4$ mag); there is rarely a question of whether you have found a Mira or not.
3. The periods are long (100 to >2000 days); although this may make them inconvenient to observe, they are ideally suited for queue-scheduled telescopes. Furthermore, the combination of period and amplitude render blending and confusion less problematic than they are for typical studies of distant populations.
4. These are late-type stars with very characteristic spectra showing molecular absorption features and emission lines at certain phases, classified as Me, Ce, Se or Ke.
5. There are a number of characteristics that are dependent on the pulsation period, so determining the period provides a good deal of information. In particular there are the period luminosity (PL) relations, discussed below. But more broadly the period tells us about the populations from which the Miras are drawn. In the Galaxy the kinematics are a function of the period and it seems likely that the same will be true of other galaxies, although little is known as yet.

AGB STARS IN GLOBULAR CLUSTERS

AGB variables in globular clusters provide a fundamental calibration for old AGB variables and are particularly important because of the independent information available for the clusters themselves e.g., distance, metallicity, etc. The PL(*K*) relation for AGB variables in Globular Clusters was discussed by Feast, Whitelock & Menzies (2002). They showed that the Mira variables in clusters with well determined distances and interstellar extinctions fell on the same PL relation as did O-rich Miras in the LMC. The SMC cluster variable, NGC121 V1, also fits this same relation. It would be useful to extend this to Miras with longer periods and work is in progress to get measurements of variables which may be members of the metal-rich clusters near the Galactic Centre, as well as to test their cluster membership. It would also be interesting to extend this work to extragalactic globular clusters, in the first instance to the known Mira in the M31 cluster G1. The C-rich Mira variables, with periods of around 500 days, found in Magellanic Cloud intermediate age clusters (Nishida et al. 2000) certainly fit on the PL(M_{bol}) relation (Whitelock et al. 2003).

There is a good deal of interest in the absolute magnitude of the tip of the AGB for comparison with extragalactic populations as it provides an estimate of the initial mass of the youngest population present among old and intermediate age stars. Because period information is not usually available, colour-magnitude diagrams are compared with those of globular clusters (Fig. 1). Our best estimate of the absolute magnitude of the tip of the AGB in Milky Way clusters is around M_K = -7.5, but the literature contains estimates as bright as M_K = -8.5 being used to show that the Miras in M31 and M32 are of comparable age to those in galactic globular clusters (Davidge 2001).

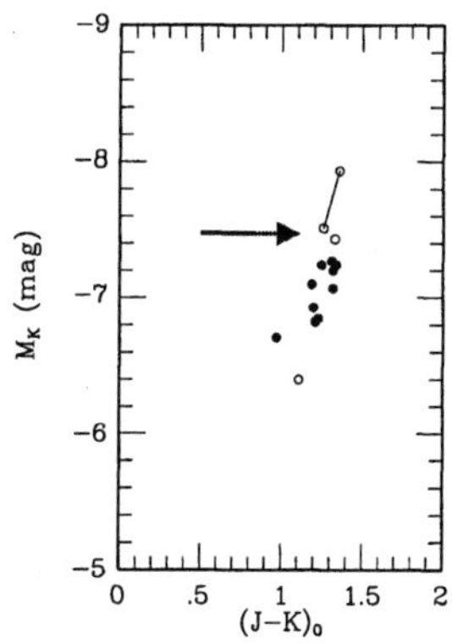

FIGURE 1. Colour-magnitude diagram for the AGB tip luminosity in globular clusters (Feast et al. 2002). The solid symbols represent well established luminosities, the open symbols uncertain ones. The arrow is the best estimate of the AGB tip luminosity for NGC 6553.

PL RELATION FOR AGB STARS

The defining work on the PL relation for AGB variables showed that O- and C-rich stars in the LMC obeyed the same PL relation at *K* (2.2μm) and a slightly different one in bolometric magnitude (Feast et al. 1989), although that slight difference may well be an artefact of the way the bolometric luminosity was determined. The scatter about the PL(*K*), σ_K=0.15 mag, is small and this relation applies to Miras with periods up to about 420 days. At longer periods the bright Miras are clearly more luminous than the PL would predict (see section on lithium below). It is the PL relation for O- and C-rich stars at near-IR wavelengths that is likely to be of primary interest with regard to the distance scale in populations beyond the Local Group.

When it became possible to examine longer period LMC Miras with very high mass-loss rates, and hence thick circumstellar shells, it became clear that they fell very close to the extension of the PL(M_{bol}) relation derived for O-rich Miras with periods under 400 days (Whitelock et al. 2003). Although these bolometrically luminous stars emit almost all their flux at mid to far-IR wavelengths, near-IR spectroscopy would still be practical with large telescopes.

What is interesting in the context of large telescopes is how we apply what we have learned from the LMC to similar stars at much larger distances, as discussed below.

AGB STARS IN THE LOCAL GROUP AND BEYOND

A great deal of work with large telescopes is in progress and the field is changing rapidly, so summaries such as this one date rapidly. We therefore highlight a few illustrative areas as indicative of work in progress and possibilities for the future. Note that Groenewegen (2004) has recently reviewed AGB stars in the Local Group and beyond, with an emphasis on C-stars which are relatively easy to identify photometrically to large distances.

NGC5128

In a very impressive piece of work Rejkuba and colleagues, using the ESO VLT, have made a detailed search for, and study of, long period variables in two fields in the halo of NGC5128 (Cen A). They obtained JHK_S colours for one epoch only, but have over 20 K_S frames per field, acquired in service mode with ISAAC (Rejkuba et al. 2003a,b), enabling the determination of good periods and mean magnitudes. They found over 1500 variables in an area of 10.46 square arcmin and determined periods and amplitudes for a large fraction of them. It is worth noting that this type of programme, which requires about 20 observations at roughly one month intervals, is ideally suited to queue-scheduled telescopes.

Almost all of the variables are above the tip of the giant branch at $K_S = 21.24$ mag and the survey is estimated to be about 50% complete for $K_S < 22.5$. Interestingly, very few of the variables have colours which might suggest they are C-stars. This is probably a consequence of the relatively high mean metallicity of the galaxy, [M/H] ~ -0.4, although it may also indicate a rather small intermediate mass population in the halo of the galaxy.

Rejkuba (2004) compared her results to the PL relation for Galactic globular clusters, selecting only variables with good periods, and colours which indicate that they are probably O-rich stars (Fig. 2). The distance modulus she deduces for NGC5128 is 27.96±0.11 mag compared to 27.87±0.16 from the tip of the giant branch. This certainly demonstrates that with good light curves and colours, which allow definite Miras with relatively thin dust shells to be selected, the PL can be very effectively applied to this type of distant population. Provided that crowding is not a problem it should be possible to apply the same techniques considerably beyond the 3.8Mpc of NGC5128.

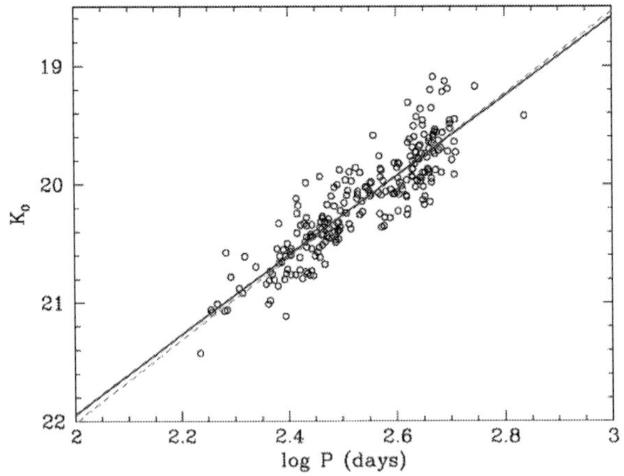

FIGURE 2. Mira PL diagram for NGC5128 from Rejkuba (2004) for variables with well determined periods and (J-K) < 1.4. The solid line is Rejkuba's best fit to the Mira sequence and the dashed line is the relation from Feast et al. (2002).

M31

A completely different approach, which opens up the possibility of studying large amplitude AGB variables in unresolved galaxies, is offered by the technique used in the POINT-AGAPE survey (An et al. 2004). This is exciting because it could be applied to much more distant galaxies as well as to the crowded bulges or disks of nearby systems. An et al. (2004) obtained regular, once per night, observations over three 6 month M31 seasons from the 2.5 m INT on La Palma. This is obviously an unnecessarily high sampling rate for AGB stars, but the primary objective was the detection and monitoring of microlensing events, so the AGB and other variable star results were just spin-off. Variability of all types was found from surface brightness fluctuations – which can provide good periods, but not magnitudes, although discrimination between bright and faint variables is possible via the amplitude of the pixel variations. They were able to conclude that the short period bright LPVs in M31 were more concentrated towards the centre of the galaxy than the fainter long period ones.

Mould, Saha & Hughes (2004) have just published a very extensive catalogue of LPVs in M31 and it will be interesting to compare their periods with those from the POINT-AGAPE survey. The Mould et al. PL diagram shows a great deal of scatter and very little period dependence. Most of the stars with periods below 400 days seem to lie above the PL relation – perhaps indicating that some of them are SR variables rather than Miras and can therefore be expected to fall near the other AGB sequences discovered by Wood (2000). There have also been measurements of small numbers of AGB variables in M31 with HST (e.g. Stephens 2004) and from large ground based telescopes (e.g. Rich, Mould & Graham 1993; Davidge 2001).

It will be interesting to see the POINT-AGAPE technique applied to other galaxies – clearly it tells us about a good deal more than just AGB stars and, again, it is very well suited to queue-scheduled telescopes.

Dwarf Spheroidal Galaxies

As discussed above in the context of NGC 5128, relating the PL relation derived from Galactic and LMC data on O-rich Miras to other galaxies works rather well. In metal-deficient populations, where the stars near the top of the AGB are usually C-rich, the problem may be more complex. Surveys of Local Group galaxies are underway and finding significant numbers of very red AGB variables, many of which will be C stars. These observations will be useful in calibrating comparable efforts at much greater distances currently underway on large telescopes (see below). For the dwarf spheroidal Leo I monitoring programs are enabling periods and mean luminosities to be determined which imply a considerable range of initial mass – much greater than limited data for the Sgr Dwarf Elliptical suggest for that system.

We (e.g. Menzies et al. 2002) are in the process of testing the PL relation as defined for LMC variables on various other Local Group galaxies. This has two main thrusts: first, as a probe of the populations within the Local Group galaxies themselves, and secondly, of more interest in the context of this meeting, to test how these methods might be applied to much more distant galaxies – where very much less is known about age, metallicity and star formation history. Fig. 3a illustrates preliminary results for Leo I. The dotted line shows where you would expect the PL to be (assuming the slope is the same as in the LMC) if the distance modulus of 22.0 mag derived from the tip of the red giant branch (Bellazzini et al. 2004) is correct. The solid line, almost 0.4 mag fainter, is the best fit to the data omitting the faint point (for reasons discussed below).

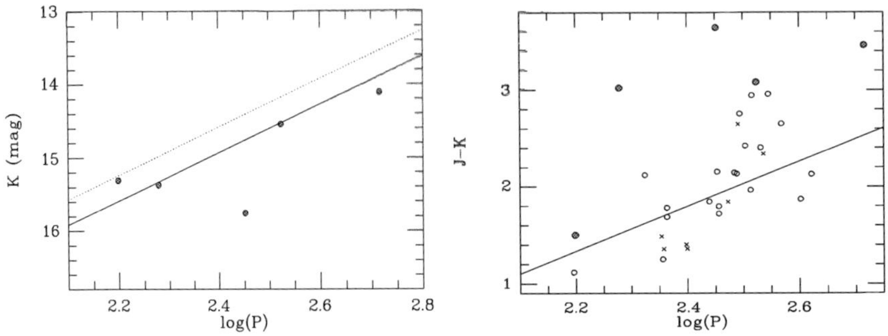

FIGURE 3. (a) PL relation for the (assumed) C-rich Miras in Leo I discussed by Menzies et al. (2002). The solid line is the best fit PL to the four most luminous points and the dotted line is the PL relation anticipated if the distance modulus is 22.0 mag; both lines are at the slope derived for Miras in the LMC. (b) Period-colour relation for the same stars, compared to C-stars in the LMC (open circles) and the Sgr dwarf elliptical (crosses).

Fig. 3b compares the colours of the 5 Miras in Leo I, to those of the LMC C-stars, which define the PL relation, and those of C-stars in the Sagittarius dwarf galaxy (Whitelock et al. 1999). There are certainly differences; the Leo I stars are in general redder than you would expect for their period (much redder stars are found in the LMC, but with periods of 500 days or more). The differences may be due to circumstellar extinction or metallicity effects of some sort, yet to be established. Indeed the large spread of colour among the LMC C-stars is not understood; it contrasts to the tight period-colour relation found for O-rich stars.

The faintest of the stars shown in Fig. 3a, which has the largest *J-K* in Fig. 3b, also shows long term trends in its light curve which are thought to be due to variable obscuration by dust, a phenomenon which is quite frequent in C-stars, but rare in single O-rich Miras (e.g. Whitelock et al. 1997). It was therefore omitted from the discussion of the PL relation above, on the grounds that we expect it to be faint. It is possible that the other stars are faint for the same reason.

Other Galaxies

The last few years have seen an increasing number of papers describing giant branch and AGB measurements within and beyond the Local Group. A vast amount of work is in progress and we can expect interesting results in the near future.

M33 has an AGB extending about 3 magnitudes from the GB tip. It has been observed with Gemini North - both in the central (Stephens & Frogel 2002) and in the outer regions (Davidge 2004). A programme of monitoring is underway (van Loon et al.) to investigate the variability of the AGB stars.

It is certainly very exciting to see the first results from adaptive optics on Gemini North. Davidge & Rigaut (2004) compare two *K′* images of M32 obtained at different epochs to deduce something about the variability of the AGB stars. For M82 a single set of *H* and *K′* images, obtained with a resolution of 0.08 arcsec, were analysed by Davidge at al. (2004). They find the tip of the Giant Branch at K'=21.7 mag and the AGB extending about 1.7 mag above this. They deduce that one AGB star, the only one with H-K' >0.5, is probably a C star. These two examples give only a flavour of what is appearing in print. They are interesting as they show what might be possible for distant galaxies with big telescopes. However, it is clear that more detailed studies will provide vastly more information – both about the galaxies and about the AGB.

KINEMATICS

Kinematics offers us an important tool in the study of AGB variables, as it does in the broader understanding of stellar populations (see also Grebel's contribution to these proceedings). Feast & Whitelock (2000) discussed the kinematics of Miras in the solar neighbourhood using Hipparcos parallaxes and radial velocities. They showed that the kinematic properties, e.g. the velocity in the direction of solar rotation, are a function of pulsation period. This is, of course, to be expected once one appreciates that the period is telling us about the initial mass of the population from which the Miras evolved. The big changes in kinematic properties occur amongst Miras with P<300d. These stars are generally thin shelled and their radial velocities could be measured optically or in the near-infrared. Rather little has been published on the kinematics of AGB stars in other galaxies; Hughes, Wood & Reid (1991) examined AGB variables in the LMC and there are a few studies of C-stars in other Local Group galaxies. A great deal of work is in progress including that by Cioni and colleagues.

Thanks to the microlensing surveys, in particular OGLE, MACHO and MOA, there is an ideal database from which to select objects for radial velocity measurement and hence for kinematic studies not just of the Miras in the LMC, but also the other AGB variable star sequences discovered by Wood (2000). For the LMC a 10m telescope is not essential to achieve the required accuracy of 1 to 2 kms^{-1}, although it would make a good project for partially cloudy conditions. Similar information will be potentially available for more distant galaxies, which will require large telescopes.

ABUNDANCES

Large telescopes offer great potential for studies of chemical abundances of GB/AGB stars, which is a broad subject of great current interest. Unfortunately, while we know a great deal about Mira variables, particularly if we can determine their periods, the complexity of the pulsating atmosphere has meant that detailed abundances remain elusive for these stars at the top of the AGB, although there is some hope that the situation will change as spectral synthesis work improves. Nevertheless, there is a considerable amount to be learned by indirect means, e.g. by studying AGB stars in situations where we know their initial abundances such as in clusters or binary systems, or from semi-quantitative measurements.

Feast & Whitelock (2000b) showed that the maximum pulsation period of the GB/AGB variables in galactic globular clusters is a function of the cluster's metallicity over at least the range -0.2>[Fe/H]>-2.0. While the longest period variables in a particular cluster are often Miras (only found in the more metal-rich clusters, i.e. those with [Fe/H]>-1.5) or their low-metallicity analogues, the SRd variables, in less populous clusters they are simply semi-regular (SR) variables, i.e. not at the tip of the AGB. We are hoping to extend this relation to higher metallicities using the obscured globular clusters near the Galactic Centre. The only extragalactic O-rich Mira for which there is comparable data is NGC121 V1, which fits on the same sequence. Large telescopes will provide data for extragalactic Miras in clusters and it will be interesting to see how they compare. The AGB variable in M31-G1 would be an obvious place to start, although there remains a great deal of uncertainty about G1's metallicity (e.g. Stephenson et al. 2001).

There is an unfortunate ambiguity in distinguishing between the effects of age and metallicity on the properties of old and intermediate age populations. A comparison of metallicity/period data for different galaxies will provide one way of examining differences in star formation history and metal enrichment.

Technetium

Technetium is of special interest as an indicator of third dredge-up, because of its relatively short half-life of $2x10^5$ yrs. Thus if Tc is detected from a particular atmosphere you know that that star must have undergone 3rd dredge-up; other s-process enhancement could be extrinsic, i.e. produced on a companion star and transferred to the one under examination. It has not to date been possible to probe the full range of AGB stars one would like to test for Tc, because the most extreme AGB stars – those with long periods and high mass-loss rates – are too faint to observe at high resolution at the critical wavelengths on medium sized telescopes. Lebzelter & Hron (1999; 2003) examined a sample of galactic variables for Tc. However, observational constraints meant that few stars with periods over 400 days were conclusively examined, and it is in these longer period stars, with their higher initial-mass, that one would expect dredge-up to be clearly present. Large telescopes with high resolution spectrographs could provide us with a real indication of when dredge-up actually starts and over what mass range it occurs.

Lithium

Lithium is the other element of special interest. Whitelock et al. (2003) suggested that the large amplitude variables in the LMC, which lie above the Mira PL(M_{bol}) relation, are actually undergoing envelope burning (also known as hot bottom burning), but the available evidence is very limited. The presence of strong lithium absorption in the spectra of AGB stars is diagnostic of envelope burning. Whitelock (2004) specifically suggested using SALT to survey for LiI 6708 Å absorption in Magellanic Cloud AGB stars with a range of properties, including those with heavy circumstellar obscuration. Combining SALT spectroscopy with infrared photometry will allow us to track the way envelope burning changes luminosities – about which there are predictions, but almost no observational evidence. It should also be possible to use SALT and HET to study lithium in AGB stars in other Local Group galaxies and thereby establish the range of initial masses of stars undergoing envelope burning and the influence of metallicity.

INTERFEROMETRY

There have been very considerable advances in the techniques of interferometry over the last few years which are already starting to impact on stellar astrophysics (see, e.g. review by Monnier 2003). There are large and small players in the field, but the VLT and Keck are already major contributors. For red giants, and AGB stars in particular, interferometric techniques will prove vital to the broad understanding of various factors, including size, shape, temperature, mass-loss, distances, magnetic fields, duplicity etc. First and foremost AGB stars are large, with typical stellar diameters of the order of 1 AU and dust shell diameters of around 1000 AU. They will be natural targets for the 1mas resolution that interferometry promises. One of the reasons we are interested in detailed interferometry is for establishing significant departures from the uniform disc (UD) model; these asymmetries may interfere with the interpretation of high precision parallax measurements anticipated from space experiments, such as GAIA. These parallaxes of Galactic Miras should eventually provide the best AGB PL calibration for distance scale measurements. Of course we are also interested in what causes any large departure from a UD – such as the atmospheric structure and details of hot or cold spots which may be related to mass-loss and magnetic fields.

Mennesson et al. (2002), using the Infrared and Optical Telescope Array, observed o Ceti – the prototypical Mira – at K' and L'; Fig. 4a shows the best UD fits for those observations, which were obtained near maximum light and therefore close to minimum diameter. Fig. 4b shows how o Ceti varies around its light cycle at L; the K variations are even larger and the light curves rarely repeat from one cycle to another. Contemporaneous interferometric measurements at different wavelengths are essential, but rarely achieved. Photometric monitoring would also aid interpretation, but is rarely performed. As can be seen in Fig 4a, o Ceti is almost 50% larger at L than at K. This seems to be fairly typical, although the difference can be as high as 100%. Critical for distance determination is the fact that diameters, even at the minimum pulsation size, are larger than parallaxes, e.g. Hipparcos measured 9 mas for the parallax of o Ceti. The measured diameter of Mira itself is at least 24 mas at K and 35 mas at L. To interpret the μas position measurements that will come from GAIA, it will be vital to understand how the illumination of the stellar disk changes around the light cycle.

Diameter measures for Miras, by occultation or by interferometry, have long been a puzzle as they very clearly indicate large radii, and thus low effective temperatures. As a consequence there has been a long standing controversy about the pulsation mode of Miras (e.g. Trimble & McFadden 1997). The low effective temperatures suggested by simple UD fits pointed to overtone pulsation, but that did not fit well with models of the shock phenomena which strongly favour fundamental pulsation. Recent evidence that the photosphere is surrounded by a circumstellar gas layer or envelope produces a good first approximation, and interferometry provides useful constraints on the envelope (Ohnaka 2004 and references therein). It is now clear that Miras pulsate in the fundamental mode and that the molecular envelope confused the diameter estimates.

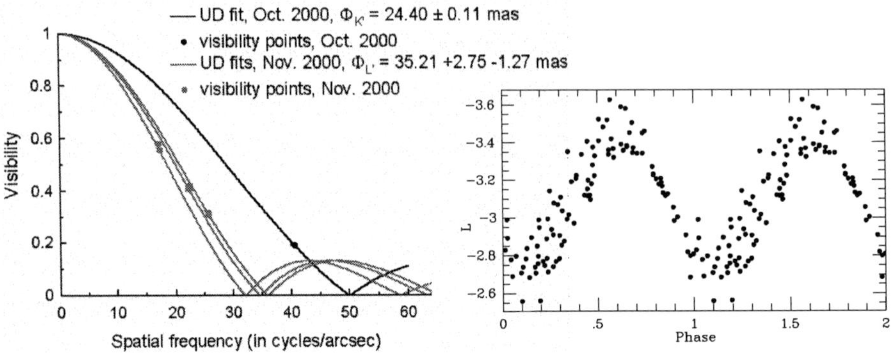

FIGURE 4. (a) Mennesson et al. (2002) observations of o Ceti in 2000 October (*K'*) and November (*L'*); Visibility measurements and best UD fits. A significant departure from the UD model is visible in the *L'* data. (b) *L* light-curve of o Ceti from Whitelock et al. (2000).

Narrow band diameters measured at different wavelengths across the *K* band, such as those obtained with the Palomar Testbed Interferometer (Thompson, Creech-Eakman & van Belle 2002), are also providing insight into the structure of the outer molecular layers. It is clear that in O-rich Miras the water absorption is very extended. This should not come as a surprise as occultation diameters of Miras at visible wavelengths indicated very different sizes in and out of the TiO bands which dominate those spectral regions (see e.g. Belega's contribution to these proceedings). It is quite clear that to get the full picture we must go to relatively narrow bands chosen strategically to measure the regions of interest – there is a wealth of information available, but extracting and interpreting it is far from trivial. In particular molecular band variations have almost certainly affected the attempts to date to measure the geometric variations in diameter that we observe as pulsation. Models suggest that the 1.04 μm pass-band is probably the region that comes closest to being true continuum (Ireland, Scholz & Wood 2004) – at least for the O-rich stars. It is also worth noting that we will only be able to make effective use of the revolution underway in interferometry if there is parallel development in theoretical modelling of AGB atmospheres.

CONCLUSIONS

In this overview we have concentrated on resolved stellar populations. However, the study of AGB stars in unresolved populations is unquestionably an important area for large telescopes, so we conclude on this topic. Fig. 5, from Maraston et al. (2001), illustrates the predicted colours of simple stellar populations with ages between about 100 Myr and a few Gyr. Three lines show theoretical loci for different metallicities, while the fourth one (below the others) shows the result of omitting the contribution from thermally pulsing AGB stars entirely – which results in very little change of *B-V*, but a huge change in *V-K*. This intermediate mass age range is the realm of the *AGB phase transition* (Renzini & Buzzoni 1986) and Fig. 5 illustrates the major contribution to light and colour from that evolutionary phase. The data points are Keck observations of young clusters in NGC7252 and they certainly appear to show the AGB phase transition contribution to their colours. NGC7252 ($(m\text{-}M)_0$=34.1 mag) is a merger remnant with a number of blue globular clusters, thought to have been formed during the merger. Interestingly, Mouhcine et al. (2002) seem to see spectroscopic signature of C-stars in W3, the most massive of these clusters. There will obviously be more of this type of work and eventually attempts to synthesize the AGB contribution to more complex population mixes. Nevertheless, the positions of these theoretical lines are strongly dependent on rather naïve assumptions about the AGB, particularly about mass loss and envelope burning. It seems likely that a full understanding of these distant populations will only be achieved if we first make use of our large telescopes to effect more detailed understanding of AGB stars in resolved populations.

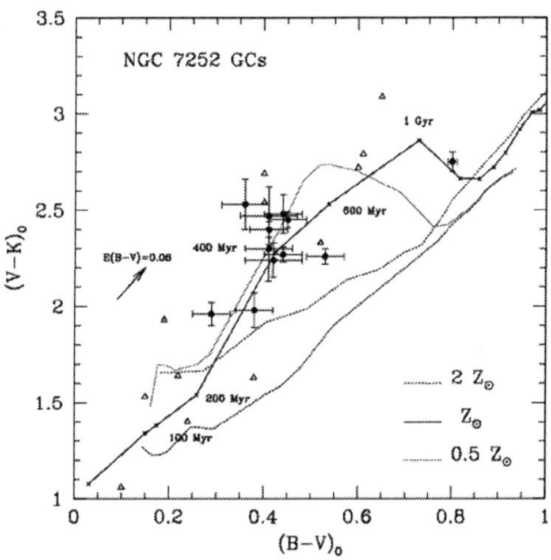

FIGURE 5. From Maraston et al. (2001) this two-colour diagram illustrates the effect of the AGB phase transition on the colours of simple stellar populations and compares them with observations of globular clusters from NGC 7252.

ACKNOWLEDGMENTS

We thank our colleagues, particularly John Menzies, for allowing us to show data in advance of publication and for a critical reading of a draft of this paper. PAW is very grateful to the organizers of this meeting for providing financial support under the auspices of the Polish-SA bilateral agreement. Martin Groenewegen is thanked for his preprint and Maria-Rosa Cioni for information on work in progress.

REFERENCES

1. Cantiello, M., Raimondo, G., Brocato, E., and Capaccioli, M., *Astr. J.* **125**, 2783-2808 (2003)
2. Bellazzini, M., Gennari, N., Ferraro, F. R., and Sollima, A., *Mon. Not. R. astr. Soc.* **379**, in press (2004).
3. Davidge, A. J., *Astr. J.* **122,** 1386-1396 (2001)
4. Davidge, A.J., *Astr. J.* **125,** 3046-3070 (2004)
5. Davidge, T. J., and Rigaut, F., *Astrophys. J.* **607**, L25-L28 (2004)
6. Davidge, T. J., Stoesz, J., Rigaut, F., Veran, J.-P., and Herriot, G., *Publs. astr. Soc., Pacif.* **116**, 1-8 (2004)
7. Feast, M., and Whitelock, P., *Mon. Not. R. astr. Soc.* **317**, 460-487 (2000a)
8. Feast, M., and Whitelock, P., *Can Mira Variables Tell us the Chemical Abundances in Stellar Systems?* in: Evolution of the Milky Way: stars versus clusters, edited by F. Matteucci and F. Giovannelli, Kluwer, 2000b, pp. 229-237
9. Feast, M. W., Glass, I. S., Whitelock, P. A., and Catchpole, R. M., *Mon. Not. R. astr. Soc.* **241**, 375-392 (1989)
10. Feast, M., Whitelock, P., and Menzies, J.W., *Mon. Not. R. astr. Soc.* **329**, L7-L12 (2002)
11. Groenewegen, M. A. T., *AGB stars in the Local Group, and beyond*, in: Planetary Nebulae beyond the Milky Way, astro-ph/0407282 (2004)
12. Habing, H. J., and Olofsson, H., *Asymptotic Giant Branch Stars*, 2003, Springer
13. Hughes, S. M. G., Wood, P. R., and Reid, N., *Astr. J.* **101**, 1304-1323 (1991)
14. Ireland, M. J., Scholz, M., and Wood, P. R., *Mon. Not. R. astr. Soc.* **352**, 318-324 (2004)
15. Lebzelter, T., and Hron, J., *Astr. & Astrophys.* **351**, 533-542 (1999)
16. Lebzelter, T., and Hron, J., *Astr. & Astrophys.* **411**, 533-542 (2003)
17. Maraston, C., Kissler-Patig, M., Brodie, J. P., Barmby, P., and Huchra, J. P., *Astr. & Astrophys.* **370**, 176-193 (2001)
18. Mennesson, B., et al., *Astrophys. J.*, **579**, 446-454 (2002)
19. Menzies, J.W., Feast, M., Tanabé, T., Whitelock, P., and Nakada, Y., *Mon. Not. R. astr. Soc.* **335**, 923-927 (2002)
20. Monnier, J.D., *Rep. Prog. Phys.* **66**, 789-857 (2003)
21. Mouhcine, M., Lançon, A., Leitherer, C., Silva, D., and Groenewegen, M. A. T., *Astr. & Astrophys.* **393**, 101-113 (2002)
22. Mould, J., Saha, A., and Hughes, S. ., *Astophys. J. Suppl.* **154,** 623-631 (2004)
23. Nishida, S., Tanabé, T., Nakada, Y., Matsumoto, S., Sekiguchi, K., and Glass, I. S., *Mon. Not. R. astr. Soc.* **313,** 136-140 (2000)
24. Ohnaka, K., *Astr. & Astrophys.* **424,** 1011-1024 (2004)
25. Rejkuba, M., Minniti, D., and Silva, D. R., *Astr. & Astrophys.* **406**, 75-85 (2003a)
26. Rejkuba, M., Minniti, D., Silva, D. R., and Bedding, T. R., *Astr. & Astrophys.* **411**, 351-360 (2003b)
27. Rejkuba, M., *Astr. & Astrophys.* **403**, 903-912 (2004)
28. Renzini, A., and Buzzoni, A., *Global properties of stellar populations and the spectral evolution of galaxies* in: Spectral Evolution of Galaxies, editors C. Chiosi, & A. Renzini, Dordecht: Reidel, 1986, pp. 195-231
29. Rich, R. M., Mould, J. R. and Graham, J. R., *Astr. J.* **106,** 2252-2280 (1993)
30. Stephens, A.W., Frogel, J. A., Freedman, W., Gallart, C., Jablonka, P., Ortolani, S., Renzini, A., Rich, R. M., and Davies, R, *Astr. J.* **121,** 2597-2609 (2001)
31. Stephens, A.W., and Frogel, J. A., *Astr. J.* **124,** 2023-2038 (2002)
32. Stephens, A.W., Frogel, J. A., DePoy, D. L., Freedman, W., Gallart, C., Jablonka, P., Renzini, A., Rich, R. M., and Davies, R., *Astr. J.* **125,** 2473-2493 (2003)
33. Thompson, R. R., Creech-Eakman, M. J., and van Belle, G. T., *Astrophys. J.,* **577**, 447-456 (2002)
34. Trimble, V., and McFadden, L. A. *Publs. astr. Soc. Pacif.* **109**, 78-113 (1997)
35. Whitelock, P. A., *Envelope Burning in Asymptotic Giant Branch Stars*, in: 2nd SALT Workshop, editor, D. Buckley, 2004, in press
36. Whitelock, P. A., Feast, M. W., Marang, F., and Overbeek, M. D., *Mon. Not. R. astr. Soc.* **288**, 512-532 (1997)
37. Whitelock, P., Menzies, J., Irwin, M., and Feast, M., "Carbon Stars in the Sagittarius Dwarf Galaxy" in *The stellar content of Local Group galaxies,* edited by P. Whitelock and R. Cannon, astr. Soc. Pacific, **192**, 1999, pp. 136-143
38. Whitelock, P. A., Marang, F., and Feast, M. W., *Mon. Not. R. astr. Soc.* **319**, 728-758 (2000)
39. Whitelock, P. A., Feast, M. W., van Loon, J. Th., and Zijlstra, A. A., *Mon. Not. R. astr. Soc.* **342**, 86-104 (2003)
40. Wood, P. R., *Proc. astr. Soc. Aust.* **17**, 18-21 (2000)

The study of planetary nebulae with large telescopes

Romano L.M. Corradi

Isaac Newton Group of Telescopes, Ap. de Correos 321, 38700 Santa Cruz de La Palma, Spain

Abstract.

Planetary nebulae (PNe) are valuable objects to discuss a number of astrophysical issues. I illustrate here two topics in which the study of PNe has become particularly fashionable in the recent years, and which are best pursued using large telescopes.

The first one concerns the use of PNe to investigate one of the less understood phenomena in the evolution of solar-type stars, namely the physics and time evolution of the strong mass loss that, at the end of the Asymptotic Giant Branch (AGB), spoils completely stars of their envelopes in only few 10^5 years.

The second one is the study of extragalactic PNe, a field that had a strong burst in the last years, as PNe have proven to be powerful test particles to discuss the cosmic distance scale, as well as the dynamics, stellar content, star formation history, and chemistry of galaxies and of the intergalactic medium.

INTRODUCTION

Planetary nebulae (PNe) are nowadays observed over a range of distances which spans 5 orders of magnitudes, from nearby Galactic nebulae, whose imaging saturates the camera of a 2-m class telescope in few seconds, to the most faraway objects found so far, at the distance of $\sim$17 Mpc of the Fornax cluster, whose study requires many hours of integration with a 10-m class telescope.

Being the evolutionary link between evolved stars with a large range of masses (1 to 8 $M_\odot$) and the interstellar medium, the study of PNe implies understanding many aspects of stellar evolution, as well as the physics, dynamics, and chemistry of the gaseous phase in all environments in the Universe.

The following discussion concentrates on two illustrative examples of modern science done with PNe, which are focussed on observational projects presently carried on, or planned for the near future, at large telescopes like the forthcoming SALT.

THE DYNAMICAL EVOLUTION OF PNE

At the end of the 80's [2] and during the following decade [22],[16], it became clear that the morphology of PNe is much more complex than previously believed. Multiple shells, bipolar outflows, complex mirror-like or point-like symmetries, and a variety of puzzling microstructures have been found in such a large fraction of nebulae that the old paradigm of a PN as a spherical bubble centred on the stellar remnant applies only to a minority of them. (Note, however, that the number of spherical PNe might be actually

CP752, *Stellar Astrophysics with the World's Largest Telescopes: First International Workshop*
edited by J. Mikołajewska and A. Olech

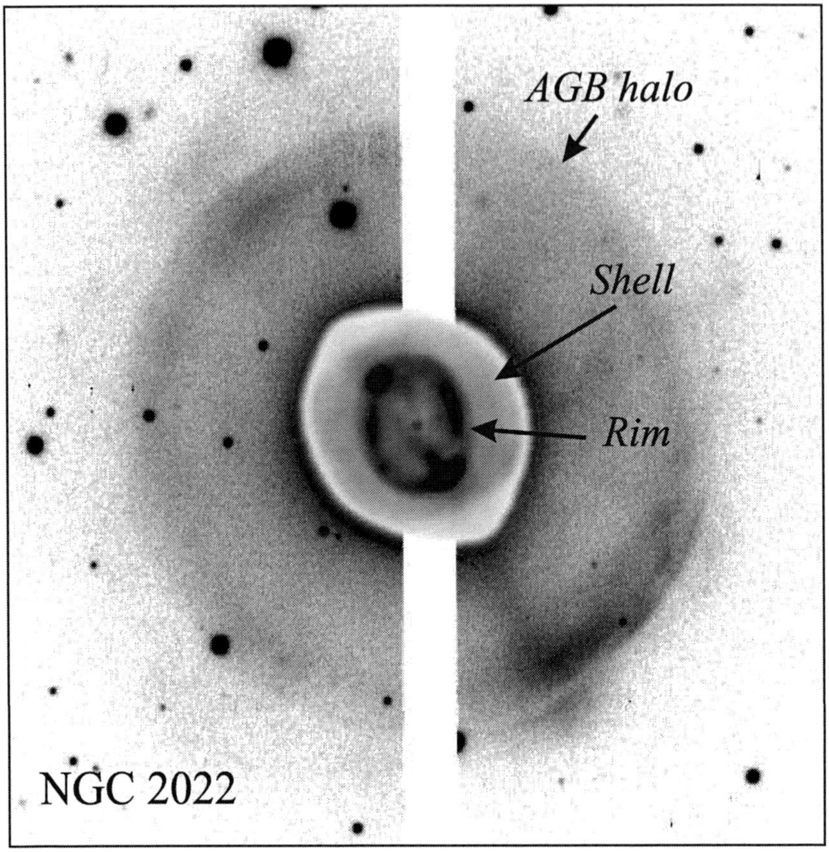

FIGURE 1. The planetary nebula NGC 2022 (from [6])

larger than observed, as a number of them have a pretty low surface brightness, and thus are more difficult to detect than other morphological types).

These observations gave a strong impulse to the dynamical modelling of PNe. The overall physical problem is both challenging and interesting, as various physical processes, relevant to the discussion of a number of astronomical fields, act simultaneously and with timescales comparable to the rapid evolution of the central star in the H-R diagram.

I concentrate here discussion on our understanding of "simple" (spherical or only slightly ellipsoidal) PNe, one example of which can be found in Fig. 1. Indicated in the figure are the three main ionized shells of NGC 2022, which can be also found in a significant fraction of Galactic PNe. These shells, from inside to outside and from the brighter to the fainter, are respectively called the *rim*, the *shell*, and the outer *halo*. The formation of these shells involves two main dynamical processes. The first one is the so-called "winds interactions", which is the hydrodynamical interaction of the different winds episodes characterizing the late AGB and early post-AGB evolution. In fact, it is believed that the stellar envelope of low and intermediate mass stars is ejected at the end of the AGB phase as a strong stellar wind lasting for several hundred thousand years. This dust-driven AGB wind is modulated by the energy changes in the stellar interiors, and specifically by the so-called *thermal pulses* (helium shell-flashes) recurring every $\sim$100 000 yr (for a solar type star) at the base of the convective hydrogen envelope. In correspondence of each thermal pulse, there is a significant increase in the mass loss rate, followed by a sudden decrease in the low luminosity phase after the pulse [4]. The evolution of this variable mass loss must to be followed by means of a detailed hydrodynamical modelling, as done for instance by [26]. After the complete ejection of the envelope at the tip of the AGB, the stellar core evolves to high temperatures at constant luminosity, and a highly supersonic, tenuous stellar wind driven by radiation pressure on resonance lines is now blown off from the central star. This post-AGB wind rapidly reaches the regions occupied by the much slower AGB wind and drives a strong shock through it. At the same time, significant dynamical effects in the structure of the AGB wind are caused by the onset of ionization, as pressure/temperature discontinuities (and consequently shock waves) are produced by ionization fronts which are moving in/out as the central star evolves in its rapid post-AGB journey through the HR diagram.

A time-dependent treatment of both ionization and winds interactions is therefore needed to model correctly the evolution of the circumstellar nebulae. Nowadays, sophisticated radiation-hydrodynamical models are available (e.g. [18],[23],[19]), which allowed us to get an overall idea of how multiple shells form in PNe like NGC 2022. Rims are believed to be matter compressed by the action of the shock produced by the fast post-AGB wind. The density structure of the shells, which generally have a surface brightness decreasing linearly with distance from the central star, is instead setup by the initially expanding ionization front, which in most cases (except for the most massive progenitors) is able to quickly ionize all the circumstellar environment [19]. Thus the density distribution through rims and shells is severely modified by the action of shocks, and preserves little memory of the original density and velocity structure at the end of the AGB. In spite of this, useful information on the density structure and temporal variation of the AGB wind can still be recovered by measuring the shell expansion velocity, as shown by [24].

The faint outer halo provides more important data on the mass loss history in the last, say, one hundred thousand years of the AGB evolution, its (usually limb-brightened) edge being the signature of the last thermal pulse on the AGB. For this reason, we call this structure as an *AGB halo* [26],[6]. The following information can be recovered by the study of AGB haloes:

- The halo's expansion velocity is a measure of the AGB wind speed.
- The dynamical age of the halo's edge tells us how many years ago the last thermal pulse occurred. Combined with the information on the age of the rim and shell, it indicates at which interpulse phase the star left the AGB, and thus the energy source (hydrogen or helium burning) active in the post-AGB phase of the stellar remnant. Most of PN central stars are found in this way to be hydrogen burners [6].
- The slope of the radial surface brightness profile of the halo can be linked to the final AGB mass loss rate.

- Detailed modelling of the halo also allows us to determine the basic chemistry of the AGB mass loss, because the dynamical evolution of the wind depends on the type of dust (e.g. carbon vs. silicates), that have different cross-section to radiation pressure and thus different acceleration properties. This result in a halo surface brightness profile that depends on its chemistry [26], a property that in principle could also be compared to direct measurements of the halo's chemical abundances.

Although obtained in an indirect way, this information is valuable as the direct study of stars at the tip of the AGB is very difficult, generally prevented by the fact that stars are shrouded by their dusty winds, and thus not directly observable (except at far-IR wavelengths). Thus AGB haloes of PNe become the key to do a sort of 'AGB paleontology', and recover information which can hardly be obtained in other way. One illustrative example is the recent discovery of "rings" (the projection in the sky of several concentric shells) in the inner regions of many PN haloes [9]. These rings indicate an (unexpected) modulation of mass loss with timescales of few hundred to few thousand years, occurring in the stellar wind only in the last 10 000 to 20 000 yr of the AGB phase, when mass loss is the largest ever in the stellar history. Considering the timescales, the formation of rings cannot be due to major changes in the stellar interior like the thermal pulses, and other mechanisms should be invoked. Rings might be produced by stellar oscillations, the action of companion stars orbiting around the dying star, the formation of waves due to a complex interplay of the coupling between dust and gas in the AGB wind, or the action of magnetic cycles somewhat similar to our own Sun's sunspot cycle.

In order to study these phenomena in detail, deep spectroscopic studies with large telescopes are needed, as AGB haloes are extremely faint, 10^3 to 10^4 times fainter than the PN rims and shells. This will be certainly a productive field of work in the next years.

Highly collimated PNe

Figure. 2 presents three examples of highly collimated structures in PNe[1], showing large-scale bipolar lobes and jets.

Compared to 'classical' elliptical shells, these outflows are characterized by expansion velocities which are one order of magnitude larger, reaching in the most extreme cases velocities as high as few thousand km s^{-1}. It is clear that powerful acceleration of the AGB wind must be provided, and a number of physical mechanisms, including stellar magnetic fields and rotation in single and binary stars, or accretion disc winds in interacting binary systems, have been proposed (see e.g. the review by [3]).

The study of these nebulae is not only interesting to understand why spherical stars switch to an extreme anisotropic mode at the end of their evolution, but is also relevant for the discussion of a number of other astrophysical fields, as the collimation mechanisms acting in these PNe are likely to be fundamentally similar to those operating in young stellar objects, nova and supernovae, and AGNs. The use of adaptive optics,

[1] There is some discussion about the classification of these objects as genuine PNe, as they might instead be the ejecta from evolved stars in interacting binary systems, like symbiotic stars.

FIGURE 2. Images of three bipolar PNe.

which will be generally available in the future in large telescopes, or even interferometry at optical wavelengths, would allow unprecedented spatial resolutions, that for nearby PNe might unveil the collimating region and resolve (with some patience) the real-time evolution of the nebulae, allowing us to determine line flows, tangential velocities and accelerations, and ultimately the forces acting on the outflowing gas particles.

EXTRAGALACTIC PNE

This is presently a very rich field of work, with many perspectives also for the coming years. The first extragalactic PNe were found in M31 in 1955 by Baade [1], but nowadays the number of (candidate) PNe found in external galaxies exceeds by one order of magnitude those known in the Milky Way (which are about 2000). This extensive search for PNe up to distances as large as some 20 Mpc has been possible thanks to the selective emission of PNe in few, narrow and characteristic emission lines. Usually, the forbidden [OIII] line at 500.7 nm is used for the search, as it is very bright; up to some 15% of the high luminosity (10^3 to 10^4 $L_\odot$) of the central star can be reprocessed into this line by the circumstellar nebula. In addition, its relatively large ionization potential allows us to distinguish, at least in a statistical sense, candidate PNe from other ionized nebulae, like for instance HII regions. An example of such an imaging search in Sextans B, a galaxy located in the outskirts of the Local Group, is shown in Fig. 3.

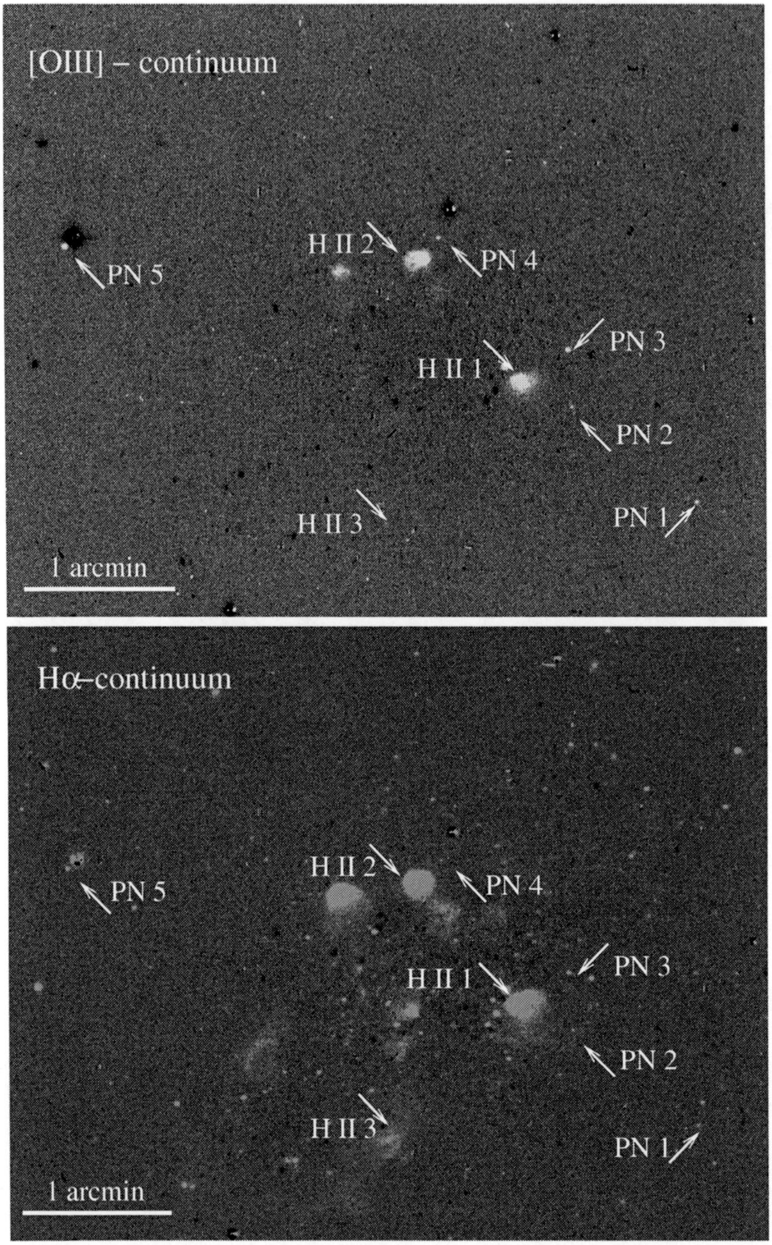

FIGURE 3. Continuum-subtracted [OIII] and Hα images of the dwarf irregular galaxy Sextans B. Emission from the underlying stellar population of the galaxy is removed by subtraction of the stellar continuum evaluated using broad band filters. PNe (labeled with the prefix 'PN') stand out as unresolved sources mainly visible in the [OIII] frames, at variance with HII regions which are usually resolved, and much brighter in the Hα image. Figure adapted from [14].

Several surveys have been conducted in the Local Group, in nearby galaxy groups and in rich clusters like Virgo and Coma, using standard narrowband imaging (followed by detailed spectroscopical studies), or special observational techniques using dedicated instruments like the Planetary Nebula Spectrograph (see below). An update reference list can be found in the proceedings of the first workshop entirely dedicated to extragalactic PNe, held at ESO Garching on May 2004 [25].

One virtue of PNe is that, being produced by stars with a large range of masses, they are found in any type of galaxy component (discs, bulges, haloes, etc.), as well as in the intergalactic space (intragroup and intracluster stellar populations). In all these environments, extragalactic PNe have proved to be excellent test particles to investigate a number of topics that are briefly discussed in the following sections, with some emphasis in the role that 10-m class telescopes are expected to play in the next years.

PNe as luminosity and age tracers

The seminal paper by [20] showed how to relate, at any given age and metallicity, the number of PNe (or of stars in any other post main-sequence phase) to the global population of a system of coeval, chemically homogeneous stars (like for instance open and globular clusters). Thus, counting PNe (which can be easily identified in distant stellar systems by their characteristic emission-line nature), directly translates into measuring the luminosity of the underlying stellar population from which they derive. In more complex systems like galaxies, which are a mixture of stellar population of different ages and metallicities, this kind of exercise is more complicated to perform, but still is possible using more sophisticated recipes. The proportionality coefficient between the number of PNe and the bolometric luminosity of the stellar system, the so-called "PN specific luminosity rate", is then predicted to depend on the galaxy type [5].

In the Local Group, we have nowadays a quite extensive knowledge of the PN population. According to the predictions of the stellar evolution theories above, a nice overall scaling relationship between the luminosity of galaxies and the number of PNe has been found [10]. However, given the dependence of the PN specific luminosity rate on the kind of stellar mixture present in each galaxy, small but non-negligible deviations from the proportionality relation is found depending on whether or not star formation at intermediate ages (2 to 8 Gyrs ago) has been particularly strong. Thus, the PN population proves to be sensitive to the relative number of stars of intermediate ages in a galaxy [15].

Some theoretical [17] and observational [11] evidence might also exist of age effects on the properties of the PN population in a galaxy, in the sense that the existence/absence of PNe with an [OIII] luminosity above a certain value might indicate the occurrence/lack of relatively strong star formation during the last $\sim$2 Gyrs.

However, it is in low surface brightness systems, where the integrated stellar light is so faint that can hardly be detected, that the quality of PNe as luminosity indicators becomes almost unbeatable. In these systems, like the extended haloes of galaxies, or the space between galaxies in groups or clusters, PNe prove to be one of the best ways to detect the presence of conspicuous stellar populations. A lot of work has been done in this field in the recent years, but much more remains to be done as it requires both to go deep and to observe large sky areas. However, very valuable results have already been obtained (see e.g. the review by [13]), and are briefly summarized in the following. About four hundred candidate PNe have been found in several regions amounting to less than 1% of the sky area covered by Virgo cluster (D$\sim$15 Mpc). Forty of these PNe have been spectroscopically confirmed. More than one hundred PNe not associated with specific galaxies have also been identified in a similar area scanned through the Fornax cluster (D$\sim$17 Mpc). These data allowed to conclude that the fraction of the total cluster luminosity that is contained in this (previously unknown) intracluster matter could be as high as 20%. This would imply an articulated dynamical history of the clusters, probably involving strong tidal stripping of stars from galaxies, and intracluster light would represent another non-negligible source of baryonic dark matter in galaxy clusters.

Quite surprisingly, a much smaller amount of intergalactic light (less than 5% of the total luminosity) is instead found in the few groups of galaxies that has been searched for, like for instance the M81 group. Why is such a difference in the amount of intergalactic light between groups and clusters of galaxies?

In any case, these preliminary results need to be improved by new imaging and spectroscopical observations, aimed at refining the estimate of the total amount of light in the intergalactic space, determining its spatial distribution and the association with tidal tails/bridges between galaxies, and equally important removing all contaminant sources, like background galaxies whose Lyman-α is redshifted to the same wavelength covered by the narrowband [OIII] filters which is used to detect extragalactic PNe (e.g. Lyman-α galaxies at $z = 3.13$ for the Virgo cluster). To date, the search for intracluster PNe has mainly been done using large field-of-view cameras at 2 to 4 m telescopes with exposure times of many hours, and this has clearly limited the coverage to very small fractions of the clusters areas. In the future, significant speed-up of the observations, with a consequent wide spatial coverage, is foreseen by the use of large telescopes, using specific techniques like for instance *tunable* filters which further improve detection by a significant factor. These tunable filters are expected to be available at telescopes like SALT, the 10m Spanish GTC telescope, and the 6.5m Magellan, the latter one also providing a large field of view (about half a degree). This would allow detection of PNe up to the considerable distance of about 50 Mpc and perhaps more. At such large distances, spectroscopic follow-up, aimed at confirming their nature as PNe and measuring their radial velocities, also definitely requires the use of 10m class telescopes.

PNe as dynamical mass tracers

Being among the few stellar representatives that can be observed in low surface brightness systems, and considering that their radial velocities can be easily measured from their emission lines, PNe also result to be excellent (discrete) tracers of the stellar dynamics in many galaxy components. They are unique test particles to probe the angular momentum distribution at large galactocentric distances in the haloes of galaxies, tidal streams, and in the intergalactic space.

The issues that can be tackled are so important that specific instruments have been designed to accomplish the goal. The Planetary Nebula Spectrograph, installed at the 4.2m WHT telescope at La Palma, uses the technique of counter-dispersed slitless spectroscopy in a narrow wavelength region around the [OIII] line, which allows to detect PNe and measure their radial velocities in one shot [12]. Applied to galaxies as far as the Virgo cluster, this instrument allowed to extend the measure of the rotation curve and velocity dispersion profile of early-type galaxies out to galactocentric distances several time larger than possible by studying absorption lines from the integrated stellar light. As an example, 3 hours exposure time at the WHT allowed to measure the radial velocity of two hundred newly discovered PNe in the prototypical elliptical galaxy NGC 3379. With these data, its dispersion profile was measured out to 6 effective radii, with the striking result that an isotropic Hernquist model at constant M/L ratio seems to be the best fitting models to the observations. This lead to the quite unexpected conclusion that little (if any) dark matter exists in the halo of NGC 3379 and other similar intermediate-luminosity elliptical galaxies [21].

Application of similar instrumentation or technique at 10m class telescopes would of course allow to perform similar studies out to even larger galactocentric distances, in more faraway galaxies, and in large areas of the intracluster space.

PNe as distance indicators

[7] proposed the [OIII] luminosity function of PNe in a galaxy as a standard candle to determine extragalactic distances. A double-exponential function is generally used to describe the PN luminosity function, and the [OIII] magnitude of its steep cutoff at bright luminosities is thought to be invariant with galaxy type and age, with a mild metallicity dependence [8]. After correcting for the metallicity effect, the consistency with the distance moduli computed using Cepheids is very good, within 1 dex for the large number of nearby galaxies considered. This makes PNe as distance indicators up to distances of some 50 Mpc using large size telescopes.

PNe as chemical tracers

PNe can also be used to discuss the chemical evolution of all galaxies in the Local Group and surroundings. In fact, the abundances of certain elements (like Ne, Ar, S, and to a first approximation also O) are not varied during the evolution of the PN progenitors, and thus the PN abundances reflect those of the interstellar medium at the time when their progenitors were born (i.e. from roughly 1 to 8 Gyr ago). This metallicity information at intermediate ages can hardly be obtained in other ways, and nicely complements the measure of the present metallicity of a galaxy which can be obtained from HII regions with the same kind of observations. On the other hand, it is also known that the progenitors of PNe produce and return to the interstellar medium elements like He, C, and N. Measuring abundances of these elements therefore allows us to test stellar evolution models and the light elements enrichment of the interstellar

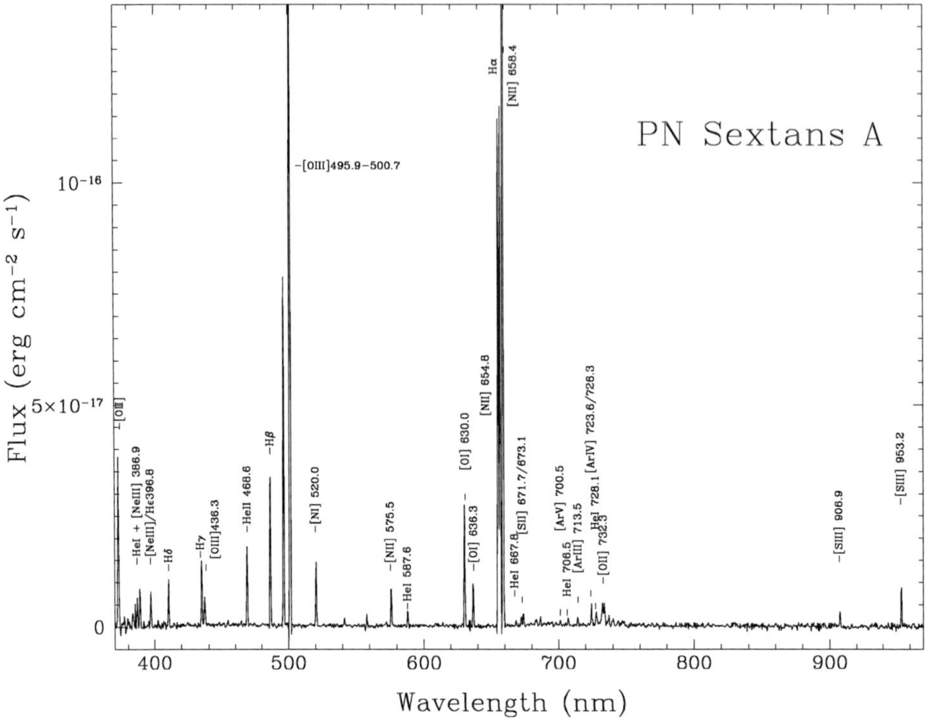

FIGURE 4. Spectrum of the only PN known in Sextans A, obtained with the multi-object spectroscopic mode of VLT+FORS2 (1.5 h in the blue side, and 1 h on the red one). Adapted from [15].

medium of a galaxy.

In order to determine accurately the chemical abundances of PNe (or HII regions), deep spectra over a large wavelength range are needed. Many studies in the literature are instead based on relatively poor spectra, and the assumptions made on the (undetermined) plasma physical conditions, in addition to other unavoidable uncertainties like the fact that only few ionization stages of each element can be generally observed, propagate into large final errors in the elemental abundances. Similarly, calibration of relationships often used to determine the chemical abundances using only few lines, are intrinsically sensitive to a number of assumptions, and in any case need to be calibrated using high quality data for a wide range of physical conditions, exciting stars, and metallicities. The most critical optical emission lines for a direct determination of the physical conditions in the gas are the electron temperature and density diagnostics, the faintest lines being [OIII]436.3 nm, [NII]575.5 nm, and the [SII]671.7,673.1 nm doublet. These can be two order of magnitude fainter than the brightest nebular lines, like Hα and [OIII]500.7 nm, and their detection requires 10m class telescopes for distances larger than 1 Mpc. An example of the observations that one wishes to obtain, and that are needed to limit the uncertainty to less than 0.3 dex in the (logarithmic) abundances of the elements mentioned at the beginning of this section, is presented in Fig. 4 (from [15]): a 2.5 h spectrum was obtained with the 8m VLT, at a resolution around 1000, for a bright PN in the low metallicity dwarf irregular galaxy Sextans A, located just outside the Local Group at a distance of 1.5 Mpc. This is the kind of observations that are needed in order to discuss details of the chemical evolution of galaxies using PNe (and HII regions), an objective that can be pursued using the large telescopes available presently (or soon) worldwide. Such detailed studies will still be limited to a distance of few Mpc, but will allow us to gain a much more complete knowledge of the chemical history of the nearby Universe.

ACKNOWLEDGMENTS

I would like to thank the organizers for such a nice conference and for the economical support provided. I am grateful to Dr. Laura Magrini for preparing Figs. 3 and 4.

REFERENCES

1. Baade, W., 1955, AJ, 60, 151
2. Balick, B., 1987, AJ, 94, 671
3. Balick, B., Frank, A., ARAA, 40, 439
4. Blöcker T., 1995 A&A, 297, 727
5. Buzzoni, A., Arnaboldi, M., Corradi, R.L.M., 2005, A&A, in preparation
6. Corradi, R.L.M., Schönberner, D., Steffen, M., & Perinotto, M. 2003, MNRAS, 340, 417
7. Ciardullo, R., Jacoby, G.H., Ford, H.C., Neill, J.D. 1989, ApJ, 339, 53
8. Ciardullo R., Feldmeier J.J., Jacoby G.H., Kuzio de Naray R., Laychak M., Durrell P.R., 2002, ApJ, 577, 31
9. Corradi, R.L.M., Sanchez-Blazquez, P., Mellema, G., Giammanco, C., Schwarz, H. E., 2004, A&A 340, 417
10. Corradi, R.L.M., Magrini, L., 2005 in "Planetary Nebulae beyond the Milky Way", ESO Astrophysics Symposia, N. Springer-Verlag, in press
11. Corradi, R.L.M., Magrini, L., Greimel, R., Irwin, M., Leisy, P., Lennon, D.J., Mampaso, A., Perinotto, M., Pollacco, D.L., Walsh, J.R., Walton, N.A., Zijlstra, A.A., 2005, A&A, in press
12. Douglas, N.G., Arnaboldi, M., Freeman, K.C., et al., 2002, PASP, 114, 1234
13. Feldmeier, J.J., 2005 in "Planetary Nebulae beyond the Milky Way", ESO Astrophysics Symposia, N. Springer-Verlag, in press
14. Magrini, L., Corradi, R.L.M., Walton, N.A., Zijlstra, A.A., Pollacco, D.L. , Walsh, J.R., Perinotto, M., Lennon, D.J., Greimel, R., 2002, A&A, 386, 869
15. Magrini, L., Leisy, P., Corradi, R.L.M., Mampaso, A., Perinotto, M., 2005, MNRAS, submitted
16. Manchado A., Guerrero M.A., Stanghellini L., & Serra-Ricart M., 1996, 'The IAC Morphological Catalog of Northern Galactic Planetary Nebulae', IAC
17. Marigo, P., Girardi, L., Weiss, A., Groenewegen, M.A.T., Chiosi, C., 2004, A&A, 423, 995
18. Mellema, G. 1994, A&A, 290, 915
19. Perinotto, M., Schönberner, D., Steffen, M., & Calonaci, C., 2004, A&A, 414, 993
20. Renzini, A., Buzzoni, A., 1986 in *Spectral Evolution of Galaxies*, eds. Chiosi C. and Renzini A., Reidel, Ap. Space Sci. Lib., Vol. 122, p.195
21. Romanowsky, A.J., Douglas, N.G., Arnaboldi, M.; Kuijken, K., Merrifield, M.R., Napolitano, N.R., Capaccioli, M., Freeman, K.C., 2003, Science, 301, 1696
22. Schwarz, H.E., Corradi, R.L.M., Melnick, J., 1992, A&AS 96, 23

23. Schönberner, D., & Steffen, M. 2003, in IAU Symp. No. 209, Planetary Nebulae: Their Evolution and Role in the Universe, ASP, p.147
24. Schönberner, D., Jacob, R., Steffen, M., Perinotto, M., Corradi, R.L.M., Acker, A., 2005, A&A, in press
25. Stanghellini, L., Walsh, J.R., Douglas, 2005, "Planetary Nebulae beyond the Milky Way", ESO Astrophysics Symposia, N. Springer-Verlag, in press
26. Steffen, M., & Schönberner, D. 2003, in IAU Symp. No. 209, Planetary Nebulae: Their Evolution and Role in the Universe, ASP, p. 439

Microstructures of Planetary Nebulae with Large Telescopes

Denise R. Gonçalves

Insituto de Astrofísica de Canarias, Tenerife, Spain
Instituto de Astronomia, Geofísica e Ciências Atmosféricas - USP, São Paulo, Brazil

Abstract. Planetary nebulae (PNe) are known to possess a variety of small-scale structures that are usually in a lower ionization state than the main body of the nebulae. The morphological and kinematic properties of these low-ionization structures (LISs) vary from type to type in the sense that LISs can appear in the form of pairs of knots, filaments, jets, and isolated features moving with velocities that either do not differ substantially from that of the ambient nebula, or instead move supersonically through the environment. The high-velocity jets and pairs of knots, also known as FLIERs, are likely to be shock-excited. So far, most of the FLIERs analyzed with ground-based small and medium telescopes, as well as with the HST, do not show the expected shock-excited features —either the bow-shock geometry or the shock excited emission lines. In this talk we discuss the crucial problem of the excitation mechanisms of FLIERs —through the comparison of jets and knots of NGC 7009 and K 4-47— and what might be the contribution of large telescopes.

MACRO- AND MICROSTRUCTURES OF PNE

The main, macro, structures of PNe are the elliptical, bipolar, or point-symmetric rims and shells, as well the haloes that dominate the optical [O III] and Hα line emission of planetary nebulae (see top panel of Figure 1). Notwithstanding important revisions, the mechanism responsable for the rim formation has been known for almost 30 years as the interacting stellar wind, ISW, models (Kwok, Purton & FitzGerald 1978). Now we believe that, in addition to the interplay between the slow AGB and the fast post-AGB winds, magnetic fields and rotation —probably due to a disk within a binary system— should also play a role in the formation of the main structures of the PNe (for a review, see Balick & Frank 2002). The formation of the shells, external to the rims, has been ascribed to the action of the photoionization front on the AGB matter not yet reached by the shock produced by the fast wind (e.g., Schönberner 2003). The outer halo in its turn is interpreted by Corradi et al. (2003) as being composed by matter ejected during the AGB phase, its outer edge marking the signature of the last AGB thermal pulse.

The microstructures of PNe, on the other hand, are far less understood theoretically. That they are much more prominent in the lower ionization optical emission lines ([N II], [S II], [O II]), appear on smaller scales than the main structures, and are usually different of the main bodies in terms of morphology and kinematics, gives us some guidelines in attempting to unveil their origin. A number of papers have recently been published on the subject, in which some aspects of their origin are treated. From observations, we know that: LISs appear as pairs of jets, knots, filaments, and jetlike features, or as isolated systems (bottom panel of Figure 1); LISs sometimes expand with the rim, shells, or haloes in which they are embedded, but sometimes they are much faster than the main

CP752, *Stellar Astrophysics with the World's Largest Telescopes: First International Workshop*
edited by J. Mikołajewska and A. Olech

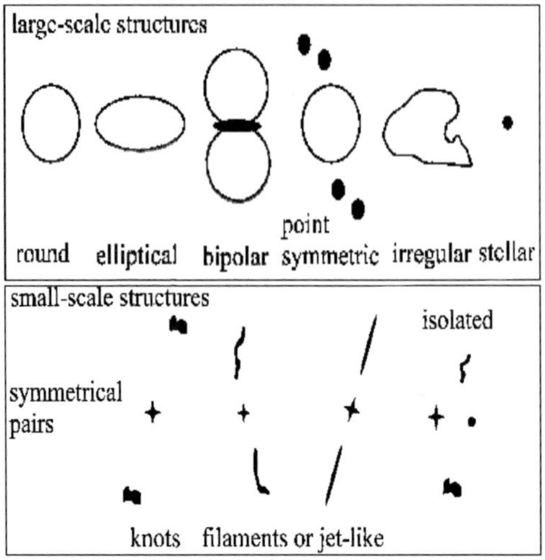

FIGURE 1. Morphological classes of PNe. Top: The main components. Bottom: The microstructures, LISs.

PNe components; they are spread indistinctly within all the morphological classes of PNe; in general, they do not have an important density contrast with respect to the main bodies; and most LIS systems studied up to now are mainly photoionized.

With regard to the comparison of LIS properties with theoretical predictions, some of their characteristics seem hard to explain (see Balick & Dwarkadas 1998; Gonçalves et al. 2001, 2003; Balick & Frank 2002). However, the origin of part of these systems (from their morphology and kinematics) can be reasonably understood via ISW models, in single stars or binaries, with or without magnetic fields, precession, and wobbling.

LOOKING FOR SHOCK EXCITATION IN LIS

Because the high-velocity pairs of knots and jets are highly supersonic, it is expected that their optical emission line spectrum show line ratios characteristic of shock-excited emission.

Figure 2 and 3 show three diagnostic diagrams for K 4-47 and NGC 7009, in which the excitation of the different regions of these PNe can be checked. From the figures, the high-velocity pair (Knot1–Knot2) of LISs in K 4-47 is mainly shock-excited, while its core is locate out of the shock regime zone, much closer to the loci of the photoionized structures. In contrast, NGC 7009 has three pairs of highly supersonic LISs —the knots K1–K4, K2–K3, and the jets J1–J2— in addition to its rim (R1–R2). In this case, none of the features, either LIS or rim, fits the zone of the shock-excited emission.

As in K 4-47, a few other PNe were found with shock-excited features: M 1-16 (Schwarz 1992, Huggins et al. 2000); and M 2-48 (López-Martín et al. 2002). What these three PNe have in common is that they are highly collimated bipolar PNe; have high-velocity structures (100 up to 300 $\mathrm{km\,s^{-1}}$); and share properties with young PNe.

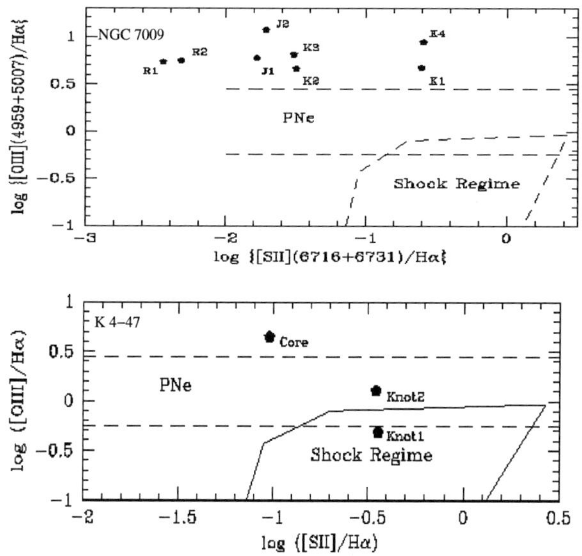

FIGURE 2. The [O III]/Hα *vs.* [S II]/Hα diagram showing the excitation of the eight regions of NGC 7009 (top) and three of K 4-47 (bottom), from 2.5 m INT long-slit medium-resolution spectra (see Gonçalves et al. 2003, 2004).

NGC 7009, on the other hand, seems to be more evolved than K 4-47, M 1-16, and M 2-4.

WHAT ABOUT "THE LARGEST TELESCOPES"?

The fact that the expected shock excitation of jets and other high-velocity LISs is not usually observed —in evolved PNe— could mean that jets/knots are relaxed systems in the sense that shock excitation is no longer present because LISs were already reached by the energetic photons of the post-AGB central star, and/or affected by local instabilities (Dopita 1997; Miranda et al. 2000; Soker & Reveg 1998). This seems to be the case for NGC 7009. In fact the PN NGC 7662 was observed with the HST (WFPC2 and STIS) and no shock-excited emission was found to be associated with its FLIERS (Perinotto et al. 2004). Therefore, it is clear that either shocks are not present in evolved PNe (not even associated with the highly supersonic LISs) or the 'the thickness of the shocked layer in LIS is too small to be resolved with HST,' as suggested by the latter authors.

Two ways of further investigating this issue are therefore: i) to determine via simulations the size of the shock zone (or the thickness of the working surface associated with the shock) of LISs in evolved PNe, and ii) observe these structures with the largest telescopes, which would give a spatial resolution better than with the HST.

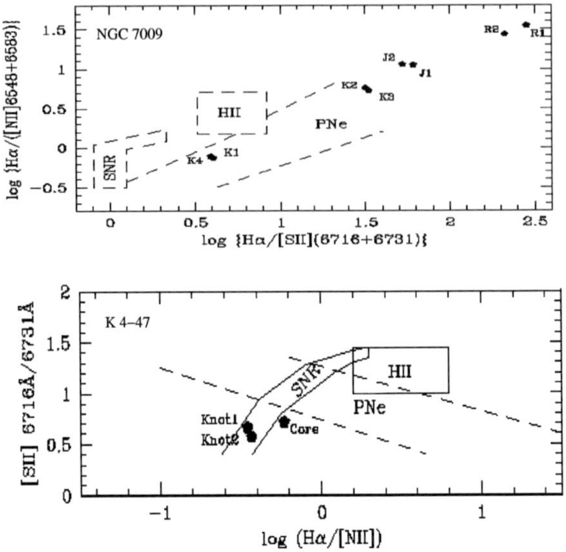

FIGURE 3. As in Fig. 2, but in the Hα/[N II] *vs.* Hα/[S II] and [S II](6716Å)/[S II](6731Å) *vs.* Hα/[N II] diagrams.

ACKNOWLEDGMENTS

I thank the Organizing Committee for the wonderful conference and acknowledge partial support from the Spanish Ministry of Science and Technology (AYA 2001-1646).

REFERENCES

1. Balick, B.; & Frank, A., 2002, ARA&A, 40, 439
2. Corradi, R. L. M, Schönberner, D., Steffen, M., Perinotto, M., 2003, MNRAS 340, 417
3. Dopita, M. A., 1997, ApJ, 485, L41
4. Gonçalves, D. R., Corradi, R. L. M., & Mampaso, A., 2001, ApJ, 547, 302
5. Gonçalves, D. R., Corradi, R. L. M., Mampaso, A., & Perinotto, M., 2003, ApJ, 597, 975
6. Gonçalves, Mampaso, A., D. R., Corradi, R. L. M., Perinotto, M., Riera, A., & López-Martín, L., 2004, MNARS 355, 37
7. Huggins, P. J., Forveille, T., Bachiller, R., & Cox, P., 2000, ApJ, 544, 889
8. Dwarkadas, V. V., & Balick, B., 1998, ApJ, 497, 267
9. Kwok, S., Purton, C. R., Fitzgerald, P.M., 1978, ApJ, 219, L125
10. López-Martín, L. et al., 2002, A&A, 388, 652
11. Miranda, L., et al., 2000, MNRAS, 311, 748
12. Perinotto, M., Patriarchi, P., Balick, B., & Corradi, R. L., 2004, A&A, 422, 963
13. Schwarz, H. E., 1992, A&A, 264, L1
14. Schönberner, D., 2003, in "Planetary Nebulae and Their Role in the Universe", proceedings of IAU Symposium No. 209, ASP Conference Series, p. 439

The Metal Abundances of Symbiotic Stars

Z.X. Zhu*, G. Zhao* and F.L.Wang*

*Purple Mountain Observatory, West Beijing Road No.2, Nanjing, 210008, P.R. China

Abstract. High resolution, high signal to noise , echelle spectra of some symbiotic stars were obtained in five runs from 1999 to 2004. The metal abundance of HD154791, which contains a neutron-star companion, is similar to other normal giants with same spectral type. The stellar surface parameters of the cool component of AG Dra have been calculated utilizing the Kurucz atmosphere model and assuming the local thermodynamic equilibrium. We have obtained the effective temperature of 3960 ± 50 K, the surface gravity of $\log(g) = 0.85 \pm 0.30$, and the metal abundance of [Fe/H]$= -1.5$. Combining the results with other authors' observations, we conclude that the cool component of AG Dra may be evolved to early AGB.

INTRODUCTION

Symbiotic stars contain interacting binary systems in which the cool component is a giant or a Mira variable and the hot companion is either a white dwarf or a neutron star or a low-mass main sequence star, and both components are embedded in an extended nebula. The hot component accretes mass from the cool companion. The gravitational potential energy provides the energy of strong emission lines which implies that a large accretion rate of the hot companion is needed to give rise to the symbiotic phenomenon, especially for systems with the hot companion being a low-mass main sequence star.

The metal abundance of the cool component of symbiotic star is an important factor, which determines the evolution including the eruption of the symbiotic and constrains the allowed models.

OBSERVATIONS AND RESULTS

AG Dra and HD154791 were observed in 1999 June, 2000 June, 2000 September, 2001 June, 2003 August and 2004 June by the 2.16-m telescope located at XingLong Station. Two or more echelle spectra were obtained each night. The coverage of the wavelength of the echelle spectra is from 5800Å to 8300Å. The signal to noise of the spectra is about 100. The resolution of the spectra is 0.25Å. The data were reduced by IRAF. Following steps have been included: subtracting the bias and the scattered light, flatten correction, extracting the spectra, wavelength calibration, normalization, etc.

HD154791

We have measured the equivalent widths of more than 40 lines of Fe I and Fe II, comparing them to the result of HR8672 (M3 III), HR7157 (M4 III), etc. There are no

CP752, *Stellar Astrophysics with the World's Largest Telescopes: First International Workshop*
edited by J. Mikołajewska and A. Olech

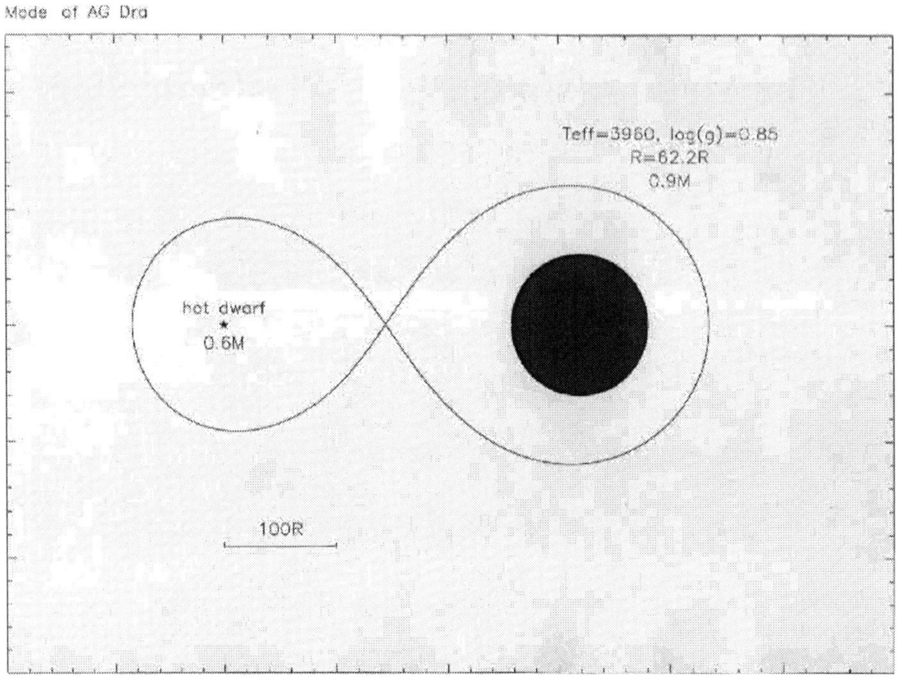

FIGURE 1. The model of AG Dra. Orbit inclination is $i = 120°$

obvious differences between them, so the metal abundance of HD154791 is similar to that of normal giants although it contains a neutron star companion.

AG Dra

More than 40 lines of Fe I and Fe II have been used to determine the stellar parameters of the cool component of AG Dra. Utilizing the Kurucz atmosphere model, based on the local thermodynamic equilibrium hypothesis, the stellar surface parameters of the cool component of AG Dra have been calculated: the effective temperature is 3960 ± 50 K, the surface gravity is $\log(g) = 0.85 \pm 0.30$, the metal abundance is [Fe/H]=-1.5.

We then use other authors' results: $f(m) = 0.0058\ M_\odot$, $a_1 \sin i = 51\ R_\odot$ (Galis et al [1]), orbit inclination $i = 120°$ ($105 - 140°$) (Schmid and Schild [2]), $M_2 = 0.6\ M_\odot$ (Mikołajewska et al. [3]). Then, the model of AG Dra could be obtained, and it is shown in Figure 1.

A star with 0.9 solar mass and 62 solar radii have evolved to its early AGB phase, when we put it on evolution trace diagram. When we change the orbit inclination, we still obtain similar results. Above result is important for the evolution and eruption of AG Dra.

CONCLUSIONS

The metal abundance of HD154791 is similar to normal giant. AG Dra is a CH star with [Fe/H]=−1.5, the cool component of it may have evolved to an early AGB star, so a high mass loss rate of the cool component of AG Dra is estimated. That will be useful to model its frequent eruptions.

REFERENCES

1. L. Galis, L. Hric, M. Friedjung and K. Petrik, *A&A*, **348**, 533 (1999).
2. H. M. Schmid and H Schild,, *A&A*, **321**, 791 (1997).
3. J. Mikolajewska, S.J. Kenyon, M. Mikolajewski, M. R. Garcia, R. S. Polidan, *AJ*, **109**, 1289 (1995).

Symbiotic Stars from the Gemini South

C. Quiroga*,†, J. Mikołajewska**, E. Brandi‡,§, O. E. Ferrer‡,† and L.G. García‡

*Facultad de Ciencias Astronómicas de la Universidad Nacional de La Plata, Paseo del Bosque S/N, 1900 La Plata, Argentina
†Consejo Nacional de Investigacioes Científicas y Técnicas (CONICET), Argentina
**N. Copernicus Astronomical Center, Warsaw, Poland
‡Facultad de Ciencias Astronómicas de la Universidad Nacional de La Plata, Argentina
§Comisión de Investigaciones Científicas de la Provincia de Buenos Aires (CIC), Argentina

Abstract. We propose to obtain high resolution infrared spectra at the 8.3 m Gemini South telescope to examine the nature of the cool components of symbiotic stars in the Magellanic Clouds. Physical parameters such as temperature, luminosity, mass loss rate and chemical abundances can provide information about their evolutionary stage and binary structure.

INTRODUCTION

Symbiotic stars are long period interacting binaries composed of a cool giant star, a very hot component and an emission nebula. The nebula is produced by ionization of the cool component wind by the hot companion. This kind of configuration makes that symbiotic stars emit over a wide wavelength range, with the hot component dominating in the far UV and the cool giant, in the infrared wavelengths.

In most cases, the hot component is a white dwarf and a few systems contain a neutron star or disk-accreting main sequence star. Among the most interesting features of symbiotic stars is their intrinsic variability. Whereas numerous spectra have been obtained in the UV to unveil the nature of the hot component, infrared spectra have not being assiduously studied. The latter is somewhat surprising, because in spite of that symbiotic stars form a very heterogeneous class of objects, their infrared colors provide a clear classification scheme: they divide into the S-type and D-type, respectively. S-types show IR spectra dominated by photospheric emission of a "normal" giant, while D-types show IR color which indicate dust emission (Webster & Allen [1]). Most of symbiotic systems ($\sim$ 80%) belong to the S-type, have orbital periods between 300 – 1300 days, and show high density nebulae which emit in mainly hydrogen and helium emission lines. On the contrary, D-type systems contain a Mira variable as the cool component, their orbital periods are naturally longer (more than 10 years) and their nebulae show lower density than the S-types. In addition, there is a small group of yellow symbiotic, with cool components of spectral type F, G, K or early R, which can be S-type or D'-type. Complete information about symbiotic stars can be found in Mikołajewska [2] and references therein.

In the Galaxy, the cool components of symbiotic stars show a strong bias towards later spectral types (M3-M6 for S-type, M6-M7 for Mira variable, respectively) when

CP752, *Stellar Astrophysics with the World's Largest Telescopes: First International Workshop*
edited by J. Mikołajewska and A. Olech

compared with red giants in the solar neighborhood (Mürset & Schmid [3]). The frequency of Mira variables is also much larger among symbiotic giants. In addition, the mass loss rates of the symbiotic giants are also systematically higher than for single giants (Mikołajewska et al. [4]). Thus, late spectral types and high mass loss are essential for triggering symbiotic activity on the white dwarf companion.

For the above mentioned reasons, knowledge of the stellar parameters of cool components in symbiotic stars is essential to understand the interaction processes and associated energetics, and to know the evolutionary stage of symbiotic stars. Unfortunately, distance and extinction estimates are often disputable, therefore it is often difficult to estimate reliable luminosity, radius and mass loss rate for the symbiotic red giants. This difficulty can be solved by analyzing the symbiotic stars in the Magellanic Clouds.

SYMBIOTIC SYSTEMS IN THE MAGELLANIC CLOUDS

The first Magellanic Cloud symbiotic star (S63) was identified by Feast & Webster [5]. At present, the number of confirmed symbiotic stars is 6 in the Small Magellanic Cloud and 8 in the Large Magellanic Cloud (Belczyński et al. [6]). For a detailed discussion of the MC symbiotic systems, we refer the reader to the papers of Morgan [7], Morgan [8], Mürset et al. [9], and Mikołajewska [10]. Although the near infrared classification into S and D-types also works for the MC systems, there are some differences as compared to the galactic symbiotic stars:

- In the Large Magellanic Cloud, four out of eight symbiotic stars are D-type, while no D-type objects were detected in the Small Magellanic Cloud (Fig. 2a of Mikołajewska [10]). This fact could be attributed to the very low metalicity in the Small Cloud. In this sense, one should remember that the dust-to-gas ratio in the outflows of obscured AGB stars depends of the inicial metallicity (Van Loon [11]).
- The spectral types of the symbiotic giants in the MCs are earlier than those of the galactic ones. However, this tendency is also followed by field stars in the MCs, and it is probably due to the lower metallicity (Mürset et al. [9]).
- According to their position in the HR diagram, symbiotic systems in the Magellanic Clouds contain low mass giants, like the galactic symbiotics. However, even in S-type symbiotic stars, the cool components are AGB giants (Mürset et al. [9], Mikołajewska [10]). In addition, this is also confirmed by the presence of C-stars in three S-type systems. That is surprising because S-type galactic systems rarely contain AGB giants. The presence of AGB cool components without dust is possibly due to the lower metallicity, i.e., too low to form enough dust in the giant envelope. In this sense, the presence of dust would be only observed at the very tip of the AGB where the mass loss is largest. It is noteworthy that the very low metalicity, galactic halo symbiotic system AG Dra falls in the same region in Fig. 2b of Mikołajewska [10] as the symbiotic stars in the SMC. However, it must be remembered that only the brightest symbiotic stars in the MCs have been so far detected.

Recently Kahabka [12] and Mikołajewska [10] have analyzed four light curves of MC symbiotic systems and for first time we know the orbital periods of three systems in the

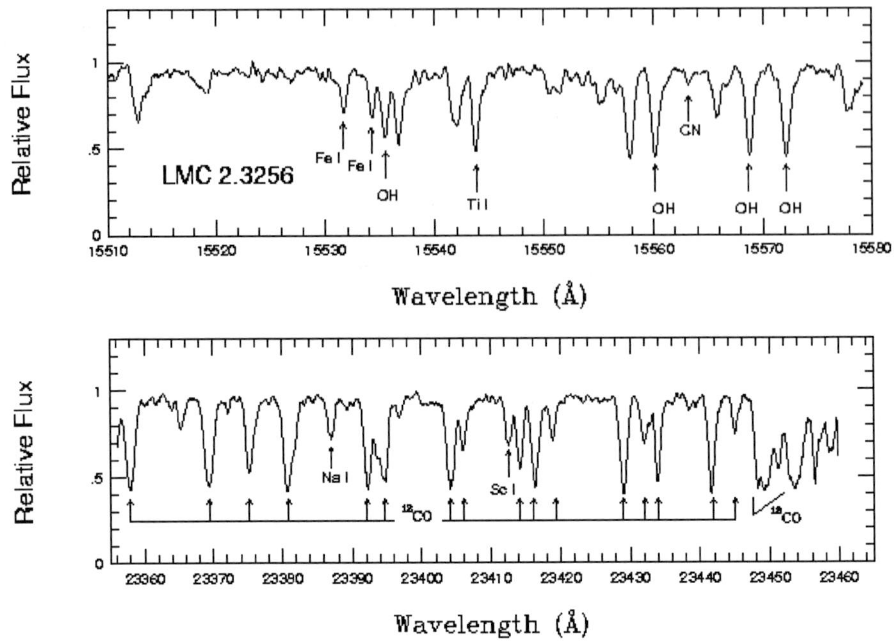

FIGURE 1. An exemplary spectrum of an M-giant in the LMC obtained by Smith et al. [13] at the Gemini South and Phoenix spectrometer.

LMC and one in the SMC. The periods found are between 900 and 1600 days, which are consistent with the longer period tail of galactic S-type symbiotic. In addition, three of these symbiotic stars show evidence for an SRa pulsations.

OUR PROPOSAL

The spectroscopic study of symbiotic stars in the infrared allows to extend our knowledge of the cool component due to the considerably lower contribution of the nebular continuum in this spectral region. With the use of a very large telescope, we can examine the nature of the cool component of symbiotic stars in the MC. For this, we propose to obtain high resolution infrared spectra at the Gemini South and Phoenix spectrometer. Fig 1 shows an example of the data that can be obtained. Twelve spectra of red giants in the Large Magellanic Cloud of similar quality were obtained by Smith et al. [13]. From them, they obtained abundances, which are in good agreement with those previously obtained for earlier spectral types.

This information, combined with the published data as well as the MACHO, OGLE and 2MASS data will provide information about:

- spectral type and luminosity of the cool components
- carbon abundances and $^{12}C/^{13}C$ from fitting CO bands with synthetic spectra.

This kind of analysis was successfully performed to obtain abundances in galactic symbiotics from K band medium resolution spectra by Schmidt & Mikołajewska [14], and in red giants in the Large Magellanic Cloud Smith et al. [13];

- iron abundances;
- rotation, turbulence and mass loss from analysis of the photospheric line profiles.

In particular, the mentioned study of abundances can provide information about the population and the past history of the system. This point is particularly interesting as there is some evidence for nonegligible metallicity effect on the symbiotic phenomenon.

ACKNOWLEDGMENTS

This study is partly supported by the Polish KBN Research Grant No. 1P03D 017 27.

REFERENCES

1. Webster, B. L., Allen, D. A., 1975, MNRAS, 171, 171
2. Mikołajewska, J. 2003, in Symbiotic Stars Probing Stellar Evolution, eds. R.L.M. Corradi, J. Mikołajewska, T. Mahoney, ASP Conf. Ser. 303, 9
3. Mürset, U., Schmid, H. M. 1999, A&AS, 137, 473
4. Mikołajewska, J., Ivison, R., R.J., Omont, A. 2002, Adv. Space Res., 30, 2045
5. Feast, M. W., Webster, B. L., 1974, MNRAS, 168, 31
6. Belczyński, K., Mikołajewska, J., Munari, U., Friedjung, M. 2000, A&ASS 146, 407
7. Morgan, D.H, 1992, MNRAS, 258, 639
8. Morgan, D.H., 1996, MNRAS, 279, 301
9. Mürset, U., Schild, H, Vogel, M., 1996, A&A, 307, 516
10. Mikołajewska, J. 2004, Rev Mex AA (Serie de Conferencias), 20, 33
11. van Loon, J Th., 2000, A&A, 354, 125
12. Kahabka, P., 2004, A&A, 416, 57
13. Smith, V.V., Hinkle, K.H., Cunha, K. et al., 2002, AJ, 124, 3241
14. Schmidt, M.R., Mikołajewska, J., in Symbiotic Stars Probing Stellar Evolution, eds. R.L.M. Corradi, J. Mikołajewska, T. Mahoney, ASP, Conf. Ser. 303, 163

Time Domain Astrophysics: X-ray Binaries in the SALT Era

P.A. Charles

SALT/SAAO, PO Box 9, Observatory 7935, Cape Town, South Africa
School of Physics & Astronomy, Univ. of Southampton, UK

Abstract.
Large ground-based telescopes allow not only the observation of very faint objects, but also allow brighter sources to be studied at high time resolution, a region of parameter space largely ignored by most giant telescope facilities. This is the field of *time domain astrophysics*, for which SALT's instrumentation will be ideally suited, and where it can make important contributions to the study of accretion processes involving white dwarfs, neutron stars and black holes. Examples of such studies of X-ray binaries and microquasars will be reviewed so as to show how their fundamental parameters and accretion disc properties can be determined.

BACKGROUND

To mention all astronomical fields in which time resolution is now a crucial factor would take an entire text in itself. Instead, after a brief introduction to the development of the technology, and a mention of the wide range of astronomical objects where time-domain investigations are important, I shall concentrate here on recent developments in interacting binaries where the nature of the physical processes occurring close to accreting compact objects provide outstanding opportunites for significant advances in our understanding, particularly at the interface of the accretion disc and compact object.

Whilst limited temporal variability studies (on necessarily long timescales) were possible with photographic plates, it was the introduction of the 1P21 photomultiplier tube in the 1940s that really marked the beginning of the modern era of the study of fast astronomical phenomena. These brought the dramatic behaviour of e.g. flare stars and cataclysmic variables (CVs) within reach of photoelectric photometry. Flickering had already been observed visually through telescopes in T CrB, AE Aqr and VV Pup, and so it was no surprise when, in the late 1940s, Linnell observed flickering of 0.01–0.2 mag on timescales of minutes or less in UX UMa (at the time, the shortest period eclipsing binary at 4.75 hrs). What was not expected was the discovery of a variable eclipse profile and the existence of a "hot spot" (a hump that lasted for roughly half the orbital cycle). Walker & Herbig (1954) perceptively suggested that this was due to a "mass of hot material situated well above the surface of the primary and located asymmetrically with respect to a line joining the two stars"!

A subsequent survey by Walker in the 1950s showed that all CVs exhibited flickering, several (probably all) were short period spectroscopic binaries, and (the most dramatic discovery from this survey) revealed precise 71.1s pulsations in DQ Her (subsequently recognised only in the 1970s as due to the strong magnetic field of the white dwarf

CP752, *Stellar Astrophysics with the World's Largest Telescopes: First International Workshop*
edited by J. Mikołajewska and A. Olech

and indicating its spin period), a result that was quite unique until twenty years later. Photometry of an outburst of SS Cyg showed that this added *non*-flickering light to the system, and also that the flickering component was blue. The next decade saw the discovery of more eclipsing systems such as U Gem. But Krzeminski's (1965) observations of U Gem (with 30s time resolution), showed that the eclipse disappeared when the source was at maximum, leading him to suggest (erroneously) that the outburst was associated with the cool secondary and not the hot primary!

It was Warner's application of next generation pulse-counting photometry (with the Texas/UCT Photometer) and a time resolution of 2s that revitalised the study of CVs. These revealed that the flickering in U Gem was located in the accretion hot spot (as the amplitude varied with the light curve) and that the disc brightens during outburst. This also explained the apparent "shift" in the eclipse timing, as it was the disc being eclipsed during outburst, but the hot spot during quiescence (a result of the somewhat lower orbital inclination). The format of the data also led to the introduction of power spectral analysis in order to search for periodic phenomena as, in most cases, the amplitudes are too weak to be visible in the raw data.

For a much more extensive historical review, together with many examples of the observations mentioned above, the reader is referred to Warner's seminal 1995 work on CVs. However, it is remarkable to note that the last decade has seen fast variability studies almost entirely concentrated in the X-ray and radio passbands, an (unintended) by-product of the (near-total) dominance of CCDs in the optical, where their long read-out times essentially eliminated the possibility of short integration times. At the time of writing, the only observatory offering common-user high-speed CCD photometry as an option at all times on its major telescopes is SAAO. The UCT CCD is the second-generation version of Warner's original fast photometer, and is based on frame-transfer capability which can reduce effective read-out times dramatically (and is built-in from day 1 on SALT; see Buckley, these proceedings).

CATACLYSMIC VARIABLES

The fundamental model for CVs of a short-period (hours to days) binary where a cool, low-mass star (usually $<1M_\odot$) transfers material via an accretion disc onto a compact object (white dwarf, neutron star or black hole) is now well established (Warner 1995 and many references therein), and we know that rapid variability is one of the key indicators of the accretion process. This variability can span a wide range of timescales as it refers to dynamical motions in the mass transfer stream and the accretion disc (hours), the spin of the compact object (tens of seconds to many minutes) and various oscillation modes (many of which are not strictly periodic). Studies of this variability can then probe the physics of the accretion disc and of the interface with the compact object itself. This latter area requires the highest time resolution, set by e.g. the gravitational free-fall timescale near the white dwarf of $\sim (G\rho)^{-1/2}$ which is of the order of seconds. The instability cycles in the accretion disc, that lead to the irregular outburst events, recur on timescales of weeks to months to years (e.g. Lasota 2001), and require long-term monitoring (e.g. by the AAVSO and RASNZVSS amateur groups) combined with TOO/over-ride status at most normal observatories.

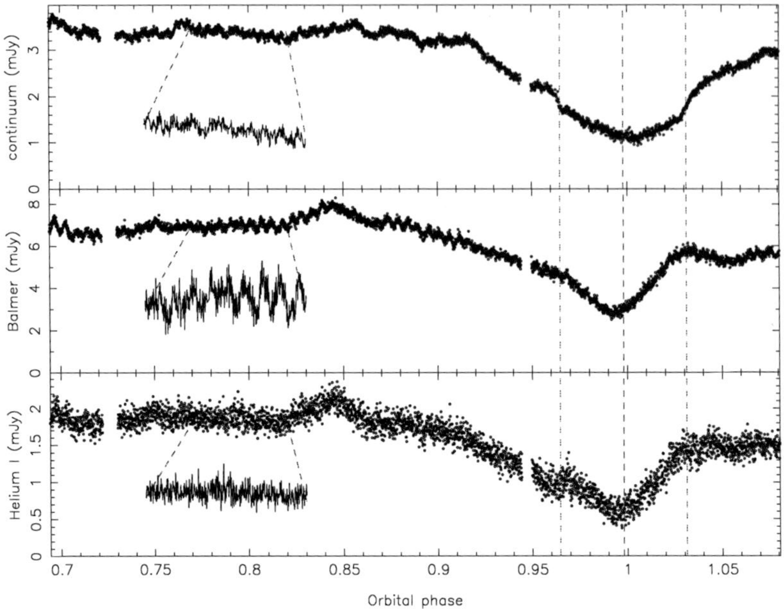

FIGURE 1. Light curves of the continuum (top), Balmer (middle) and HeI (bottom) line emission from V2051 Oph as observed with the Keck II telescope at a time resolution of 0.5s. Zoomed regions show the 56s pulsations visible in the continuum, and the 29.8s pulsations in the emission lines. The interval when the white dwarf is eclipsed is also marked. 0.1 in phase corresponds to 9 mins. (From Steeghs et al 2001.)

Dwarf Nova Oscillations

The nature of the *dwarf nova oscillations* (DNOs; in the range 10-30s) in CVs has remained a puzzle in spite of their discovery decades ago. These are not always present, not usually coherent and the periods can change even in the same object, but their values imply an origin close to the white dwarf. In spite of many models being proposed (see e.g. Warner 1995), none can account for all their observational properties. But there has been renewed interest in them as possible analogues of the much faster *quasi-periodic oscillations* (QPOs) in low-mass X-ray binaries (LMXBs), which are believed to be related to the interaction of the rapidly spinning magnetosphere of the neutron star with the inner accretion disc.

However, to investigate such phenomena, even in the brightest CVs, requires a time resolution of $\leq$1s and hence use of the largest telescopes. Consequently Steeghs et al (2001) utilised a custom-built CCD setup with the low-resolution spectrograph on Keck II to obtain high speed spectrophotometry of the eclipsing (90 min period) dwarf nova V2051 Oph. A sample of their data (in 0.5s bins) is shown in figure 1, where the 56s oscillations in the continuum (together with its first harmonic) and $\sim$29s oscillations in the emission lines are clearly visible. In fact the emission lines show coherent oscillations at 29.77s and 28.05s, plus an amplitude modulation at 8 mins (the beat period).

Steeghs et al have proposed that 56s is the spin period of the white dwarf, but it contains *two* equatorial hot spots (where it accretes from the disc) which then illuminate the disc. The disc has a substantial bulge at the circularisation radius for the system (at 12 R_{WD}), the Keplerian period of which is 488s, hence producing the twin oscillations in the emission lines. In fact, there is considerably more information contained in these data about the geometry of the disc, which is fully described by Steeghs et al. But the determination of this structure was only possible through a remarkable combination of kinematic (i.e. spectral) and temporal information, which had never been possible before.

Interestingly, such "double-frequency oscillations" had been seen before with HST during a superoutburst of OY Car (Marsh & Horne 1998), this time at around 18s, one of which they suggest is again the spin period of the white dwarf. They also speculate on reprocessing or Keplerian explanations for the second oscillation. However, an important link with LMXBs has been proposed for the DNO phenomenon by Warner et al (2003) who point out that such behaviour can now be extrapolated to explain the double kHz QPOs seen in neutron star LMXBs and, at slower frequencies, in black-hole systems. Hence the physical interpretation of this phenomenon has implications over a much wider field.

Doppler Tomography

The detailed structure of the disc can be revealed when such data are processed into Doppler tomograms (see Marsh 2001), an exceptionally powerful technique that converts the temporal resolution of the spectra into effective angular resolutions at the ($\sim$100pc distant) CVs of $\sim$micro-arcseconds. In this way an early outburst of the dwarf nova IP Peg was observed by Steeghs et al (1997) which showed the development and evolution of spiral density waves through the disc (see figure 2). Such structure has now been described by theoretical models of the disc outburst, and the observations are finally beginning to constrain the detailed physics of the increased transmission of material through the disc.

LOW-MASS X-RAY BINARIES

Low-mass X-ray binaries (LMXBs) are the much rarer analogue of CVs in which the white dwarf is replaced by either a neutron star or black hole. The much deeper potential well of the compact object leads to much greater X-ray luminosities, up to $L_X \sim 10^{38}$erg s^{-1} (the Eddington limit for a typical neutron star) and beyond (see e.g. the many detailed reviews of their behaviour in Lewin et al 1995). This means that the optical emission from LMXBs is completely dominated by reprocessed X-radiation from the accretion disc and (sometimes) from the heated face of the companion star. This opens up the prospect of using correlations between the X-ray and optical emission to probe the binary geometry. If fast X-ray variability (e.g. bursts or pulsations) is present then the time signature of the delay between them contains information about the location of the emitting regions. This is known as "echo-mapping" (O'Brien & Horne 2001) and is a technology still in its infancy in terms of how much we have learnt about LMXBs,

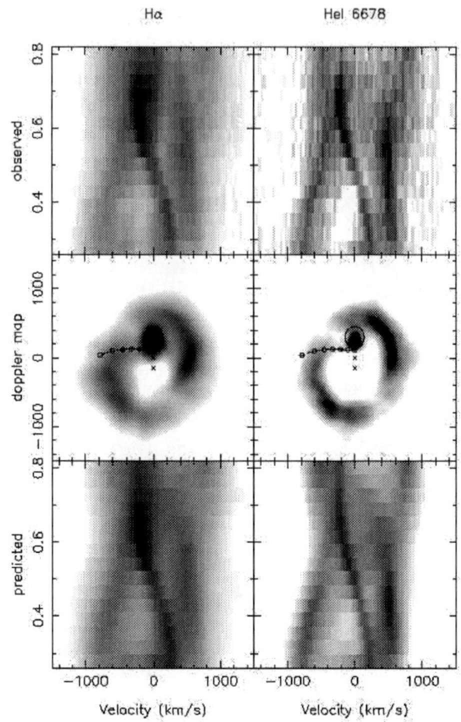

FIGURE 2. Top panels: time-resolved emission line profiles of Hα and HeIλ6678 obtained early during an outburst of IP Peg; Middle: the doppler tomograms reveal the spiral wave structure in the disc (positions of the secondary and mass transfer stream in velocity coordinates are marked); Bottom: predicted line profile variations for the calculated maps. (From Steeghs 2001).

largely due to the limited ground-based telescopes on which such necessarily fast timing observations can be undertaken (the light travel time across a typical 5-hr period LMXB is only 4-5 secs), coupled with the logistical difficulties of arranging simultaneous X-ray observations. Nevertheless, time-delayed echoes have been reported in black-hole X-ray transients in outburst (e.g. Hynes et al 1998) and, on much longer timescales, in AGN (Peterson et al 1998; Horne 2001).

X-ray Bursts

Shortly after the discovery of type I X-ray bursts in the mid-1970s, campaigns were undertaken to observe simultaneous bursts over X-ray and optical wavelengths in order to study X-ray reprocessing in detail. The logistical difficulties in such work were amply demonstrated in their first campaign (1978) by the fact that not a single burst was detected simultaneously because (a) the occurrences of bursts were unpredictable, (b) only conventional photoelectric photometers were available with the required time resolution and (c) the ground-based observations were subject to the normal vagaries of the weather! Nevertheless their persistence paid off and in 1979 and 1980 they were successful (see e.g. Pedersen et al 1982). Indeed, delays between the X-ray and optical bursts were found, and were typically ~2–3 secs (see fig 3, an example of a burst observed with a modern fast CCD), but also demonstrated a smearing effect as the X-ray burst radiation was reprocessed in the accretion disc. No reprocessed burst signature has yet been seen from the companion star (partly due to the natural shadowing of the companion by the accretion disc itself).

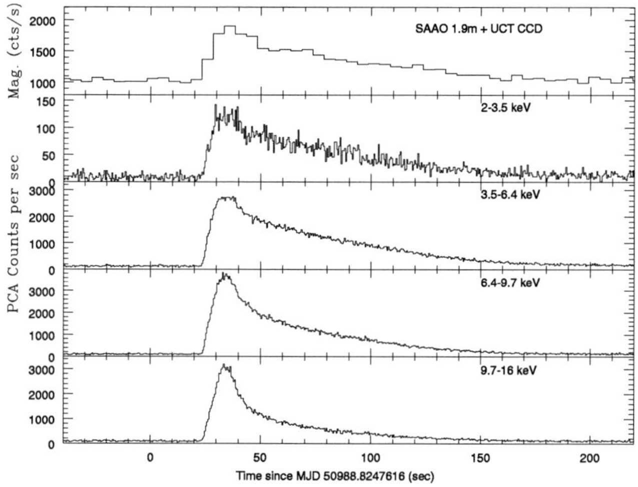

FIGURE 3. Simultaneous X-ray (RXTE) and optical (SAAO 1.9m + UCT/CCD) observations of a type I X-ray burst from the "clockwork" burster, GS1826-24 (from Kong et al 2000).

4U1626-67: the 41 min binary

The low-mass donor star in LMXBs suggests that they can have long life-times and hence the neutron stars will be old, with presumably weak magnetic fields. This accounts for the very small number of X-ray pulsars amongst LMXBs, with 4U1626-67 being for many years the only example, with a 7s X-ray pulsation. Its optical counterpart, KZ TrA (a V=18.5 blue star) was found to exhibit a corresponding (~3% amplitude) visible pulsation, but it required the much greater sensitivity of a 4m telescope (using only conventional photomultiplier tubes) for Middleditch et al (1981) to discover the much weaker (~0.5%) sideband to this pulsation and hence reveal the 41 min orbital period (of this almost face-on system, with $i \sim 18°$). As a demonstration of the power of modern, fast (frame-transfer) CCDs in this field (an early version of the type to be employed in SALTICAM and PFIS on SALT), I show in figure 4, the power spectrum of MXB1626-67 as derived from a single ~8-hr time-series with the SAAO 1.9m telescope (Chakrabarty et al 2001). This is actually the first independent confirmation of the 41 min period, and also reveals the wide range of variability present in the source. The

48mHz (P~20s) QPO is driven by an X-ray modulation at the same frequency, and is presumed to arise via an interaction between the disc and magnetosphere. The 1mHz QPO, however, has *no* X-ray analogue, and hence must be a property of the disc itself, most likely the inner regions.

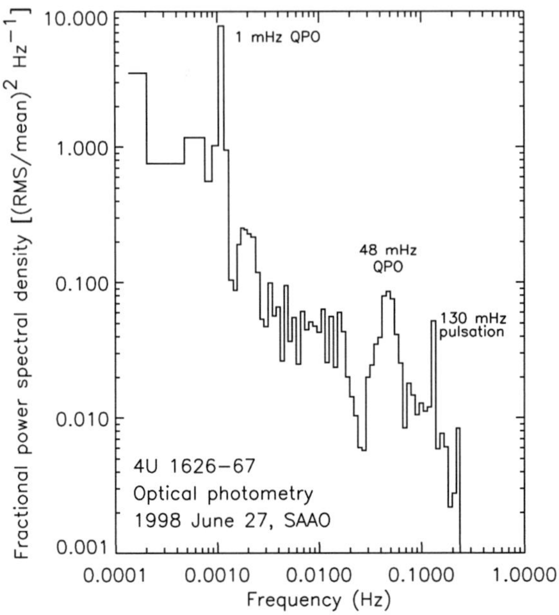

FIGURE 4. Power spectrum of an unbroken ~8-hr optical time series of 4U1626-67 obtained with the SAAO 1.9m + UCT/CCD in white light. The X-ray/optical pulsation at 7s (130mHz) is clearly seen, together with a 48mHz and 1mHz QPO, the latter of which has no X-ray analogue. (From Chakrabarty et al 2001.)

Warped, precessing discs

That accretion discs are frequently not the stable, co-planar constructs of most artists' impressions has actually been known from the relatively early days of X-ray astronomy through the 35 day on-off cycle of Her X-1. With its 1.7 day X-ray eclipses and optical modulation providing crucial geometrical constraints, it was recognised by Petterson (1975) and Gerend & Boynton (1976) that the 35 day cycle must be due to the precession of a tilted disc which periodically blocks our view of the neutron star. This is now widely accepted but the physical details little understood (see e.g. Wijers & Pringle 1999; Ogilvie & Dubus 2001), although it is usually attributed to the effects of X-ray irradiation of the disc.

More recently the high inclination X-ray dipper, A1916-05, has been shown to exhibit behaviour that is even more complicated, suggesting that the disc is actually warped and precessing. A1916-05 exhibits X-ray dips with a period of 50.00 mins, but the optical analogue of this has a period of 50.46 mins, a difference of only 1%, but now highly significant. The shortness of this period already indicates that this LMXB must be a *double-degenerate*, in that the mass donor is also (at least a partially) degenerate star (H deficient). A campaign of simultaneous X-ray (RXTE) and optical (NOT) photometry over a period of a week (chosen so as to sample the beat frequency between the two periods) revealed that the light curve evolved dramatically over this timescale (see fig 5, from Homer et al 2001), eventually repeating (although curiously not at the exact beat period, the reason for which is not understood). Similar behaviour has now also been seen in MS1603+2600 (Hakala et al 1998; 2004).

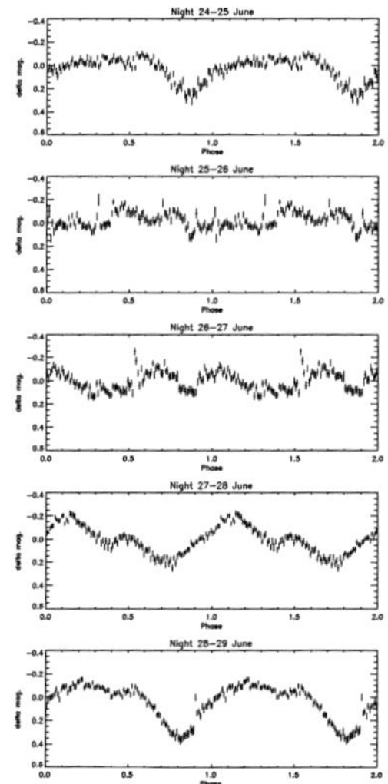

FIGURE 5. Folded optical light curves of A1916-05 obtained with the 2.5m NOT on La Palma (operating in white light) on a sequence of nights in June 1998. The substantial evolution of the light curve repeats on a timescale of $\sim$5 days (from Homer et al 2001).

Since the X-ray dipping is produced by variable azimuthal structure in the edge of the accretion disc, systems at these very particular inclination angles are very important, because any regular change in height of this structure (e.g. in a warp) will produce an additional modulation in the X-ray and optical behaviour, again allowing temporal behaviour to be mapped into geometrical information. Combined with its short period (which allows many cycles to be observed in convenient intervals) and low mass secondary, A1916-05 must be of very high mass ratio ($q = M_X/M_{sec} \geq 15$) and hence susceptible to the precessing disc instability that is seen as superhumps in SU UMa systems (see e.g. Warner 1995). But in A1916-05 we see evidence for warps as well.

However, to study these dips requires time resolution of <10s and, since the optical counterpart has V=21, to obtain anything more than the white-light photometry shown here demands much larger telescopes! Not only is SALT ideal for such systems, it more importantly has the queue-scheduled facility that makes it ideal for following the longer-term evolution of the light curve that holds the key to investigating the physics of the response of accretion discs to X-ray irradiation, again with the potential of application on a much wider astrophysical stage.

X-ray Transients

Perhaps the ultimate class of source for which a queue-scheduled telescope like SALT is ideal for studying is the X-ray transients, both in their high-mass (i.e. Be X-ray binaries) and low-mass (LMXB) variants. Charles & Coe (2003) have extensively reviewed both types, but it is the LMXB sub-class, known as *X-ray novae* (XRN) or *soft X-ray transients* (SXTs), which has proven to be of great interest recently because of the possibility of direct detection of the mass donor once the system has entered X-ray quiescence. This allows for both time-resolved photometry (which reveals the ellipsoidal modulation of the secondary) and spectroscopy (which yields the mass function). Combining these it is even possible to obtain a full solution of the system masses, and in this way more than a dozen compact object masses have been determined. Remarkably, the XRN appear dominated by black-hole candidates (BHC), where $M_X \geq 3M_\odot$, and the distribution of masses is of importance for understanding the late stages of evolution of massive stars (see Casares 2004).

The XRN which are BHC necessarily have extremely high mass ratios as well, which means that, during outburst, the accretion disc should suffer the same precessional instability as is seen in A1916-05 and the SU UMa systems. The existence of superhumps was first confirmed by O'Donoghue & Charles (1996), but the period difference (between superhump and orbital period) becomes smaller as the mass ratio increases. This makes it much harder to determine observationally as the beat period is longer and hence requires a much longer observing baseline, as beautifully demonstrated by Zurita et al (2002) from months of photometry of XTE J1118+480 during its outburst decline phase (fig 6). The period differential is so small, that the beat period is ~52d! But such studies allow an indication of dynamical properties, since theoretical models (also shown in fig 6) show a dependence of this period differential on q, which is supported by the observations.

Finally, a new method of determining kinematic information of the mass donor on both luminous X-ray binaries and XRN during outburst has been developed recently following the seminal paper on Sco X-1 by Steeghs & Casares (2002). In spite of its being the brightest LMXB (optically and in X-rays), no sign of the mass donor had ever been seen, because the optical emission is totally dominated by the X-ray-illuminated accretion disc. But the first high resolution spectroscopy of Sco X-1 (fig 7) revealed sharp components superposed on the broad Bowen fluorescence complex at $\lambda\lambda$4640-50, and these were found to be moving in anti-phase with the broad HeII emission, which originates in the X-ray illuminated, hot inner disc regions. Hence the sharp components must arise in the heated face of the donor, and contain dynamical information

This not only opens up the possibility of determining the mass functions of as many as 20 of the luminous, "steady" galactic LMXBs (a program that is well underway), it can also be applied to XRN during their outburst phase, provided that TOO/over-ride programs are in place. In the case of SALT, this will be perfectly straightforward since all observations are of necessity queue-scheduled. An example of what can be achieved

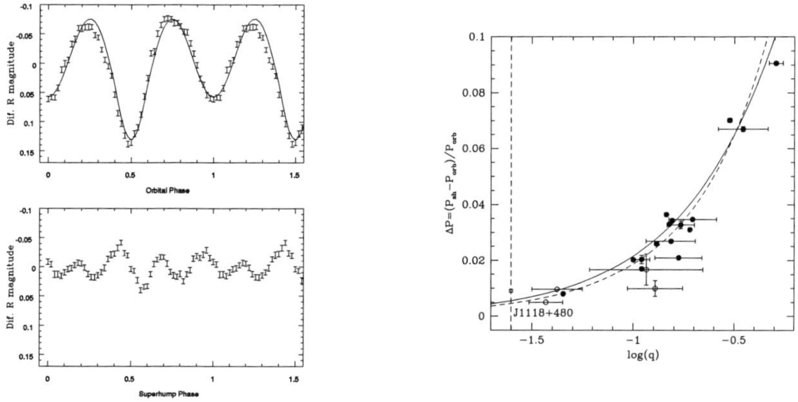

FIGURE 6. *Left*: Mean lightcurve (upper) of J1118+480 folded on P_{orb}=4.078h, together with the best-fitting ellipsoidal model of i=75° and q=25, the residuals from which are plotted below after folding on P_{sup}=4.092 hours. *Right*: Period difference, Δ, is here plotted against q for both LMXBs (J1118+480, X1916-053) and CVs (Patterson 2001) that display superhumps. Theoretical models are plotted as solid (Patterson) and dashed (Osaki 1985) curves, but these should not be valid below $q = 0.025$ (Zurita et al 2002).

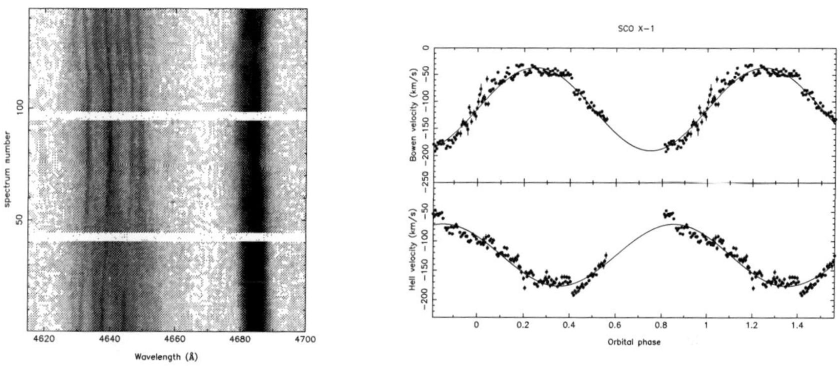

FIGURE 7. Time-resolved spectra (*left*) and radial velocity curves (*right*) of the Bowen/HeII region of Sco X-1 (Steeghs & Casares 2002).

was demonstrated by Hynes et al (2003) when GX339-4 (long thought to be a BHC because of its X-ray properties) turned on in X-rays after an extended quiescent interval. TOO spectra with the AAT, NTT and VLT indeed revealed sharp Bowen components that, finally, indicated the true binary period (1.76d) and gave $f(M)$=5.8±0.5$M_\odot$, hence confirming its BH status.

CONCLUSIONS

This brief and incomplete survey of fast astronomical photometry and spectroscopy can provide little more than a whetting of the appetite for the huge range of astrophysical phenomena that are finally becoming accessible to modern high efficiency detectors that offer the highest time resolution. Many of these areas are being greatly enhanced with the advent of the new era of giant telescopes which allow much fainter objects to be studied on shorter timescales. But it must also be recognised that much of the work of time-domain astrophysics is accessible with and requires the support of smaller telescopes and the much longer observing allocations that are possible. The combination of these with very fast detectors on giant telescopes such as SALT will truly open a new era in time-domain astrophysics.

ACKNOWLEDGMENTS

I am particularly grateful to Jorge Casares, Remon Cornelisse and Danny Steeghs for their assistance in preparing figures for this review. I also thank the organisers of this workshop for their generous support and warm hospitality.

REFERENCES

1. Chakrabarty, D. et al 2001, *Ap.J.*, 562, 985
2. Charles, P.A. & Coe, M.J. 2003, to appear in "Compact Stellar X-ray Sources", eds Lewin & van der Klis, CUP, astro-ph/0308020
3. Gerend, D. & Boynton, P. 1976, *Ap.J.*, 209, 562
4. Hakala, P. et al 1998, *A&A*, 333, 540
5. Hakala, P. et al 2004, *MNRAS* (in press), astro-ph/0410606
6. Homer, L. et al 2001, *MNRAS*, 322, 827
7. Horne, K. 2001, in "Astrotomography", eds Boffin, Steeghs, Cuypers, Springer, p386
8. Hynes, R.I. et al 1998, *MNRAS*, 299, L37
9. Hynes, R.I. et al 2003, *Ap.J.*, 583, L95
10. Krzeminski, W. 1965, *Ap.J.*, 142, 1051
11. Kong, A.K.H. et al 2001, *MNRAS*, 311, 405
12. Lasota, J.-P. 2001, *New Astron.Rev.*, 45, 449
13. Lewin, W.H.G. et al 1995, in "X-ray Binaries", CUP
14. Linnell, A.P. 1950, Harvard Circ. No. 455
15. Marsh, T.R. & Horne, K. 1998, *MNRAS*, 299, 921
16. Marsh, T.R. 2001, in "Astrotomography", eds Boffin, Steeghs, Cuypers, Springer, p1
17. Middleditch, J. et al 1981, *Ap.J.*, 244, 1001
18. O'Brien, K. & Horne, K. 2001, in "Astrotomography", eds Boffin, Steeghs, Cuypers, Springer, p416
19. Ogilvie, G.I. & Dubus, G. 2001, *MNRAS*, 320, 485
20. Osaki, Y. 1985, *A&A*, 144, 369
21. Patterson, J. 2001, *PASP*, 113, 736
22. Pedersen, H. et al 1982, *Ap.J.*, 263, 325
23. Peterson, B.M. et al 1998, *PASP*, 110, 660
24. Petterson, J.A. 1975, *Ap.J.*, 201, L61
25. Steeghs, D. et al 1997, *MNRAS*, 290, L28
26. Steeghs, D. 2001, in "Astrotomography", eds Boffin, Steeghs, Cuypers, Springer, p45
27. Steeghs, D. et al 2001, *MNRAS*, 323, 484
28. Steeghs, D. & Casares, J. 2002, *ApJ*, 568, 273

29. Walker, M.F. 1956, *Ap.J.*, 123, 68
30. Walker, M.F. & Herbig, G. 1954, *Ap.J.*, 120, 278
31. Warner, B. 1995, "Cataclysmic Variables", CUP
32. Warner, B. 2001, in "The Physics of Cataclysmic Variables and related objects", Göttingen
33. Wijers, R.A.M.J. & Pringle, J.E. 1999, *MNRAS*, 308, 207
34. Zurita, C. et al 2002, *MNRAS*, 333, 791

Gamma-Ray Bursts in the SALT/Swift Era: GRB/SN Connection

Krzysztof Z. Stanek

kstanek@cfa.harvard.edu
Harvard-Smithsonian Center for Astrophysics
60 Garden St., MS20
Cambridge, MA 02138
USA

Abstract. I discuss the Gamma Ray Burst research in the era of SALT and Swift, concentrating on the GRB/SN connection.

INTRODUCTION

The mechanism that produces gamma-ray bursts (GRBs) has been the subject of considerable speculation during the three decades since their discovery (Klebesadel, Strong & Olson 1973). The CGRO and BATSE observations demonstrated that GRBs are isotropically distributed on the sky (Meegan et al. 1992), which could have been explained with either a Galactic (Lamb 1995) or a cosmological (Paczyński 1995) spatial distribution. The BeppoSAX satellite (Boella et al. 1997) contributed the breakthrough in this field by providing rapid and accurate localizations of X-ray afterglows of GRBs. Such precision allowed for quick optical (Groot et al. 1997; van Paradijs et al. 1997) and radio (Frail et al. 1997) identifications of transients associated with individual GRBs. The identification of the optical transient (OT) associated with GRB 970508 led to the first optical spectroscopic redshift determination for a GRB, placing it at $z \geq 0.835$ (Metzger et al. 1997), and thus firmly at a cosmological distance. However, despite the great progress made in GRB science since 1997, many problems remain open. I discuss some of them, concentrating on the GRB/SN connection. I also discuss some of the exciting GRB science that could be done with the SALT telescope.

GRB SCIENCE WITH LARGE TELESCOPES

Large telescopes played a very significant role in the GRB science since the discovery of first GRB afterglows in 1997. Obviously, large telescopes can do things that smaller telescopes cannot do, such as obtaining spectroscopy of very faint sources (such as faint afterglows or GRB host galaxies) or high-resolution spectroscopy of fairly faint objects. Perhaps more importantly, given the unpredictable timing of GRBs, large telescopes allow us to obtain very significant results in *relatively short* telescope time, defined as "how much telescope time can I ask for from a random (even if friendly) observer on a remote telescope who has her/his own important program to do, before he/she gets

CP752, *Stellar Astrophysics with the World's Largest Telescopes: First International Workshop*
edited by J. Mikołajewska and A. Olech

annoyed with me"? It is my experience that up to one hour per night is doable. It is important to realize that most large telescope users do not sit around waiting for a GRB worker to call them and tell them what to do, so common sense should be applied. Common Polish heritage and passion for pierogies helps in some cases. Of course, with a queue-scheduled telescope like SALT, the life of a GRB researcher becomes a bit easier, but not boring.

Below I list several areas of important GRB science explored with the largest telescopes:

- Redshifts of the GRBs: This can be measured via absorption lines in an intervening galaxy or galaxies (as was done for the first GRB redshift by Metzger et al. 1997 using Keck 10-m telescope), via emission lines of the host galaxy superimposed on the featureless power-law spectrum of the afterglow (as was done for example for GRB 011121 by Garnavich et al. 2003a using Magellan 6.5-m telescope), or via absorption lines in the spectrum of the host galaxy. For a review on the hosts of GRBs, their redshifts and properties see Djorgovski et al. (2003) The first method, in principle, does not require large telescopes, given a bright enough afterglow and the presence of absorption lines (indeed, Jha et al. 2001 were able to measure three different absorption systems in the afterglow of GRB 010222 using our FLWO 1.5-m telescope!), but large aperture helps, given that most afterglows decay quickly.
- High-resolution spectroscopy of GRB afterglows: High-resolution spectra have now been obtained for several afterglows (see for example Schaefer et al. 2003). In a number of cases a complicated line structure with multiple components has been observed (see for example a nice paper by Mirabal et al. 2003 on "GRB 021004: A Possible Shell Nebula around a Wolf-Rayet Star Gamma-Ray Burst Progenitor"), allowing us to probe the enviroment of the GRB.
- Optical polarization of the GRB afterglows: This is truly the domain of the largest telescopes, given the photon-starved nature of polarization research (i.e. trying to measure a very small effect in the light of fairly faint objects). Despite this challenge, polarization has been now successfully measured in a number of afterglows. The first detection was accomplished using imaging polarimetry of a fairly bright afterglow of GRB 990510 with the 8-m VLT (Covino et al. 1999; Wijers et al. 1999). Spectropolarimetry has been obtained for several afterglows now as well, starting with GRB 020813 (Barth et al. 2003), using Keck 10-m telescope. The level of optical polarization is usually low, 1-2%, except in the case of GRB 020405, where 9.9% polarization has been measured 1.3 days after the burst (Bersier et al. 2003a) with the 6.5-m MMT.

The above was not intended as a review, but rather as a brief illustration of some of the many interesting results obtained for the GRBs with the largest telescopes. Needless to say, many exciting results for the GRBs were obtained with mid-sized or in some cases truly small telescopes (9th magnitude optical flash from the GRB 990123, observed with a 10-cm robotic ROTSE telescope, comes to mind: Akerlof et al. 1999). For an interested reader, there are many excellent reviews out there to learn more about GRBs. I found the recent review by Piran (2004) to be very useful.

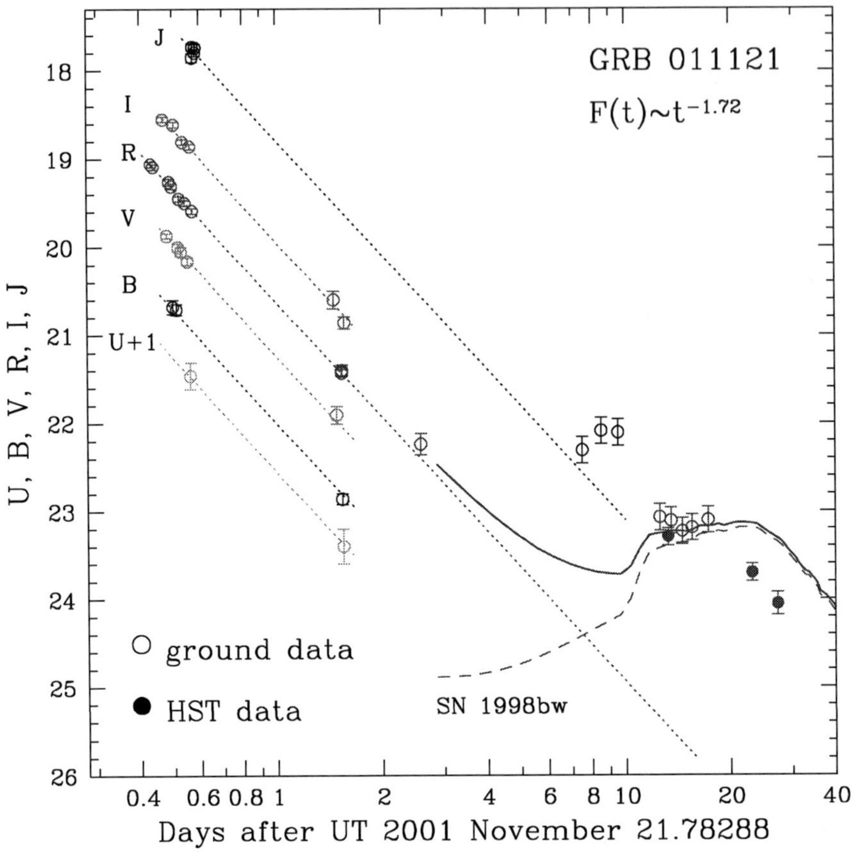

FIGURE 1. *UBVRIJ* light curves of GRB 011121. Included are three *HST* F702W epochs converted to the standard *R*-band. Dotted lines show the OT power-law decay and the dashed line is the light curve of hypernova SN 1998bw redshifted to $z = 0.36$, converted to the *R*-band, corrected for extinction and scaled by 0.1 mag. The solid line shows the combination of the OA and SN 1998bw (from Garnavich et al. 2003a).

GAMMA-RAY BURST/SUPERNOVA CONNECTION

The measured redshift of a typical GRB is $z \approx 1$, implying that a supernova component underlying an optical afterglow would be difficult to detect. At $z \approx 1$, even a bright core-collapse event would peak at $R > 23$ mag. Nevertheless, late-time deviations from the power-law decline typically observed for optical afterglows have been seen and these bumps in the light curves have been interpreted as evidence for supernovae (for a recent summary, see Bloom 2003). GRB 980425 was likely associated with "hypernova" 1998bw (Galama et al. 1998), but the isotropic energy of that burst was 10^{-3} to 10^{-4}

times weaker than classical cosmological GRBs which placed it in a unique class. Before March 2003, the best evidence that classical, long-duration gamma-ray bursts are generated by core-collapse supernovae was provided by GRB 011121. It was at $z = 0.36$, so the supernova component would have been relatively bright. A bump in the light curve was observed both from the ground and with *HST* (Garnavich et al. 2003; Bloom et al. 2002). The color changes in the light curve of GRB 011121 were also consistent with a supernova (designated SN 2001ke), but a spectrum obtained by Garnavich et al. (2003) during the time that the bump was apparent did not show any features that could be definitively identified as originating from a supernova.

While SN 1998bw was a strong hint of the GRB-SN association, no optical afterglow was observed. Without that direct association, the link between GRBs and SNe was still in question. The 'monster burst' of 2003, GRB 030329 provided that link. The burst was extremely bright in gamma-rays, implying that it was relatively close, as was quickly confirmed by the VLT spectroscopy of the afterglow, which yielded $z = 0.168$ (Greiner et al. 2003). As the afterglow faded, subtle features appeared in the normally flat power-law spectrum of the afterglow. By subtracting a continuum based upon the early shape of the spectrum, this structure was revealed as the spectrum of an unusual Type Ic SN similar to SN 1998bw, designated SN 2003dh (Matheson et al. 2003a; Garnavich et al. 2003b; Stanek et al. 2003). Within a few days, the SN became the dominant component in the spectrum (Stanek et al. 2003; Kawabata et al. 2003; Hjorth et al. 2003w).

Using the early power-law continuum spectrum as a model, one could decompose the observed spectra at later times into two separate components: GRB afterglow and SN spectrum. Using a least-squares technique, the best match for the SN among the low-redshift sample was SN 1998bw (Matheson et al. 2003b). In fact, taking into account cosmological time dilation, the spectroscopic evolution of SN 2003dh almost exactly matched SN 1998bw. Models of these spectra are presented by Mazzali et al. (2003).

An important point about the appearance of the SN was that the light curve did not show the bump that is characteristic of a rebrightening caused by the SN (see Matheson et al. 2003b and Lipkin et al. 2004 for a discussion of the light curve, which actually showed many bumps). Without the spectroscopic confirmation, the presence of SN in GRB 030329 would still be argued about.

Nebular-phase spectra of SN 2003dh show a spectrum similar to a typical Type Ic SN. Kosugi et al. (2004) present a spectrum at an age of $\sim$3 months. A spectrum obtained with the Keck telescope by Filippenko, Chornock, & Foley (2004) in December of 2003 is much like a normal Type Ic SN (Bersier et al. 2005, in preparation).

Following the discovery of SN 2003dh, reexamination of spectra of an earlier burst yielded some evidence for a SN component. Della Valle et al. (2003) found that a spectrum of the very faint afterglow of GRB 021211 had structure similar to an SN. In this case, however, the SN did not match SN 1998bw or any other peculiar Type Ic SN, but it was most similar to SN 1994I, a relatively normal Type Ic.

Another example of the GRB/SN connection came with GRB 031203. Despite high foreground reddening, spectroscopy with the VLT revealed an SN component, designated SN 2003lw (Malesani et al. 2004). For this SN, SN 1998bw was again a good match. Of the four SNe with clear GRB associations, three show remarkably similar spectra.

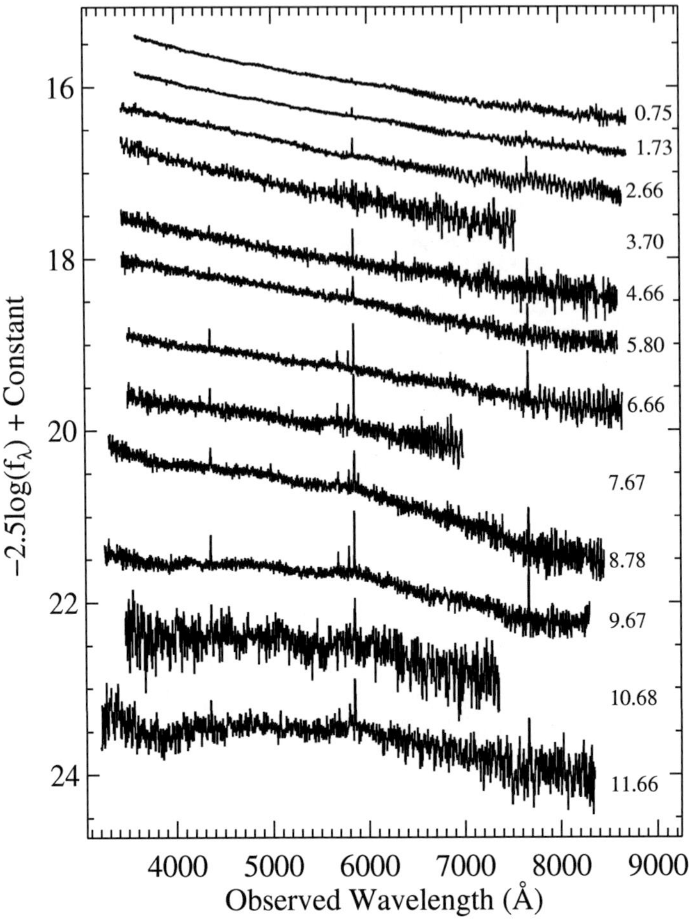

FIGURE 2. Evolution of the GRB 030329/SN 2003dh spectrum, from March 30.23 UT (0.75 days after the burst), to April 10.14 UT (11.66 days after the burst). The early spectra consist of a power-law continuum with narrow emission lines originating from H II regions in the host galaxy at $z = 0.1685$. Spectra taken after $\Delta T = 6.66$ days show the development of broad peaks characteristic of a supernova (from Matheson et al. 2003b).

FUTURE DIRECTIONS OF THE GRB/SN CONNECTION

There is now indisputable evidence that some classical (long duration/soft spectrum) GRBs come from core-collapse supernovae. But there are only two classical bursts (GRB 030329 and GRB 031203) where the progenitor supernovae have been definitively classified. SN 2003dh and SN 2003lw were type Ic events similar to the progenitor of the underluminous GRB 980425. Some theorists have concluded that all supernovae that

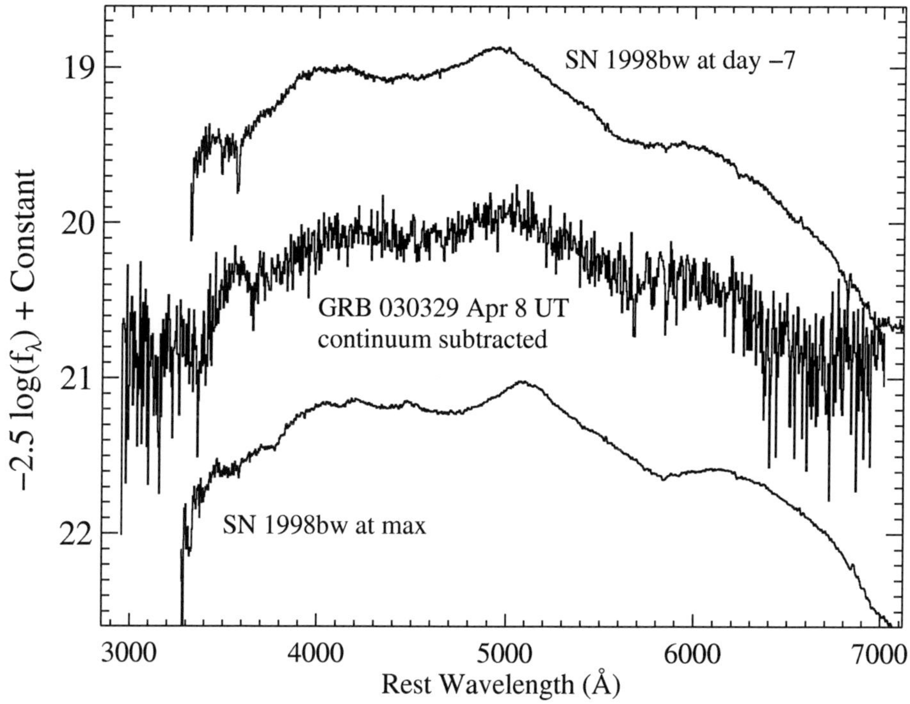

FIGURE 3. MMT spectrum of GRB 030329 from April 8 with the smoothed MMT spectrum of April 1 scaled and subtracted. The residual spectrum shows broad bumps at approximately 5000Å and 4200Å (rest frame), which is similar to the spectrum of the peculiar type Ic SN 1998bw a week before maximum light. The match is not as good for SN 1998bw at maximum light (from Stanek et al. 2003).

produce GRB are type Ic. The bias toward type Ic is partly due to the perceived difficulty in getting the jet to escape from a star with a massive envelope. But this prejudice may not be justified; after all, it was recently believed that no supernova could possibly be a GRB source because of the large baryon content of supernovae. I feel that the range of SN types that are responsible for GRBs is an unsolved observational problem.

There are hints of a diversity in the supernova progenitors of GRB. SN1997cy was a powerful type II supernova that may have been associated with a classical GRB (Germany et al. 2000), though connection to a particular GRB has not been strongly established. A better indication comes from GRB 011121 and its associated supernova 2001ke. Garnavich et al. (2003a) found the color near maximum of SN 2001ke was very

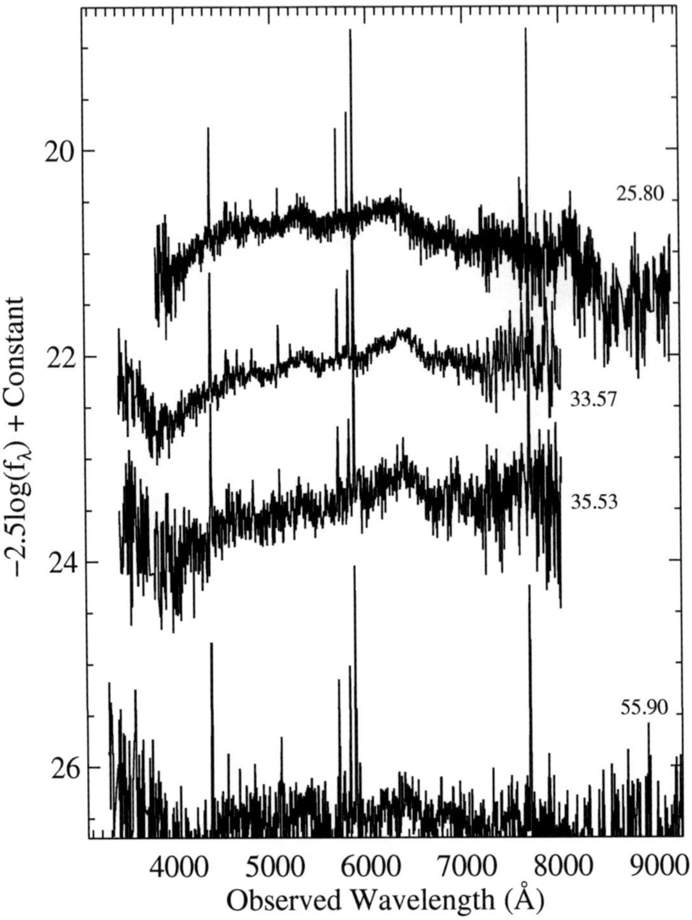

FIGURE 4. Evolution of the GRB 030329/SN 2003dh spectrum, from April 24.28 UT (25.8 days after the burst), to May 24.38 (55.9 days after the burst). The power-law contribution decreases and the spectra become more red as the SN component begins to dominate. The upturn at blue wavelengths may still be the power law. The broad features of a supernova are readily apparent, and the overall spectrum continues to resemble that of SN 1998bw several days after maximum (from Matheson et al. 2003b).

blue which is more consistent with a type II event than a type Ic. Unfortunately, there was no spectrum of the event with sufficient quality to establish the supernova type.

The small number of supernovae directly associated with GRB also leaves the question of the energetics unanswered. The supernovae that produced GRB 980425 and 030329 were two of the most energetic core-collapse events recorded. But the supernova that produced GRB 011121 was somewhat fainter and faded quickly. What is the range of supernova luminosities? Is there a connection between the energy of core col-

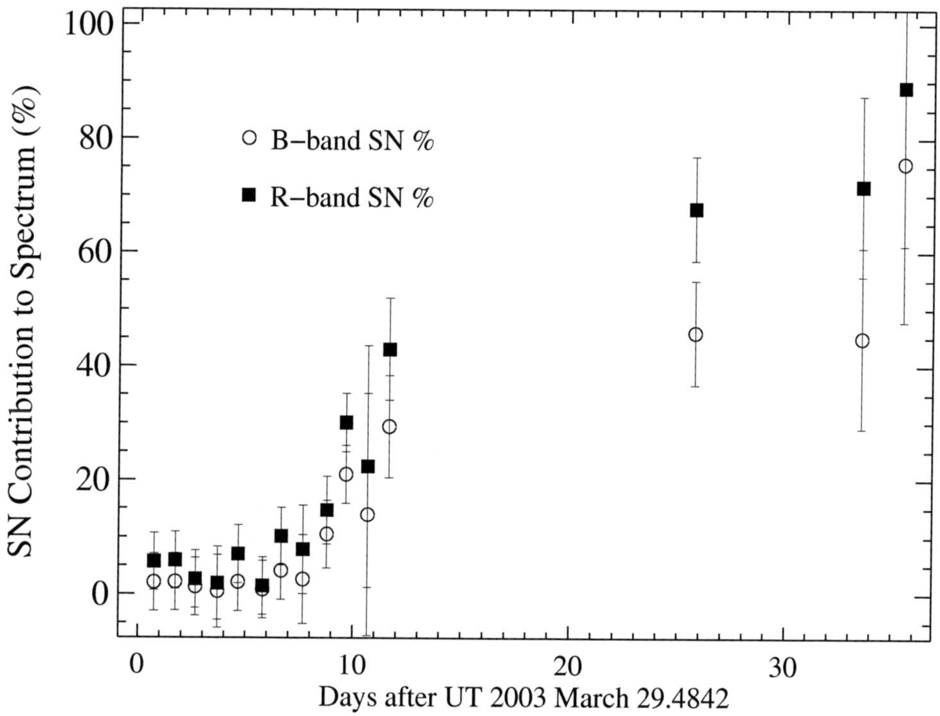

FIGURE 5. Relative contribution of a supernova spectrum to the GRB 030329/SN 2003dh afterglow as a function of time in the B (*open circles*) and R (*filled squares*) bands. We derived a best fit to the afterglow spectrum at each epoch with the fiducial power-law continuum and the closest match from our set of peculiar SNe Ic. We then synthesize the relative B-band and R-band contributions. There is some scatter for the early epochs due to noise in the spectra, but a clear deviation is evident starting at $\Delta T = 7.67$ days, with a subsequent rapid increase in the fraction of the overall spectrum contributed by the SN. Errors are estimated from the scatter when the SN component is close to zero ($\Delta T < 6$ days) and from the scale of the error in the least-squares minimization (from Matheson et al. 2003b).

lapse and the GRB properties? Frail et al. (2001) corrected the observed GRB energy for the beam opening angle and found that most bursts produce a total energy of 10^{51} erg. While this is a very large amount of energy, it is still an order of magnitude less than the total energy in a typical core-collapse supernova. Clearly, there is sufficient energy in a core-collapse to allow a wide range of supernovae to produce high-energy beams. The greatly improved sensitivity of *Swift* will allow low-energy bursts to be detectable out to a large volume of space. This will be an important test of the single-energy reservoir theory as well as the ability of supernovae with a range of properties to create GRB.

The few supernovae associated with GRBs also lead theorists to conclude that GRB are produced promptly by the core-collapse of massive stars. But this could be an observational bias. It is possible that GRBs occur within a wide range of times following

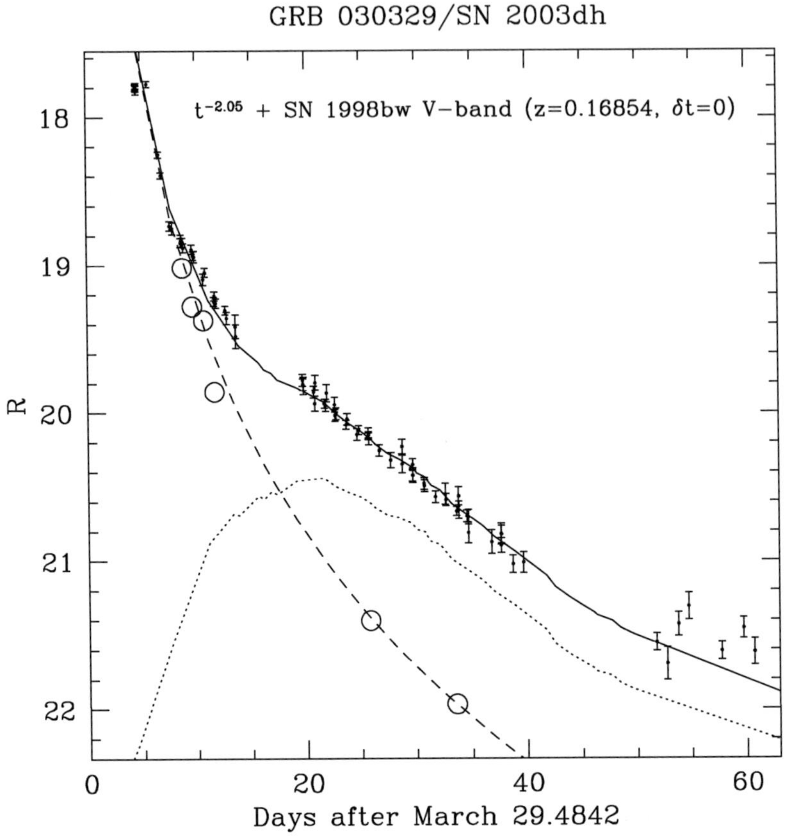

FIGURE 6. Decomposition of the GRB 030329 OT R-band light curve into the supernova (*dotted line*) and the power-law continuum (*dashed line*). As the light curve model for the supernova, we took the V-band light curve of SN 1998bw stretched by $(1+z) = 1.1685$. The resulting supernova light curve peaks at $m_R = 20.4$ (from Matheson et al. 2003b).

core-collapse, and we only detected those few bursts that occurred less than a few days after the supernova. In support of this possibility, a number of GRB have not been seen with accompanying supernovae (e.g. Price et al. 2003). The 'supranova' model predicts that the final collapse of a neutron star into a black hole can be delayed by seconds to years after the initial core-collapse. While it would be difficult to show a supernova went off years before a burst using optical wavelengths, supernovae that exploded weeks to a month before the GRB could be detected with careful observation.

Obtaining magnitudes, colors and spectra of more GRB supernovae is clearly a top priority in understanding the origin of long/soft bursts. Since the typical GRB has a redshift of $z \sim 1$, this work requires large telescopes such as SALT.

A MODEST PROPOSAL FOR SALT IN THE SWIFT ERA

The Swift satellite (Gehrels et al. 2004) will push the GRB research into an even higher gear, with > 100 GRBs per year rapidly and accurately localized. SALT is very well positioned to take full advantage of Swift, due to a number of factors:

- Unique position on the globe;
- Queue schedule;
- Stable instrumentation for several years;

For the description of SALT and its capabilities, see Buckley 2004 (this volume). The first item on the list is obvious, i.e. despite limitations on pointing, due to its location in South Africa SALT will be able to access some of the GRBs not possible to observe from the North or to observe them earlier than possible from Chile, where other large southern telescopes are located. Queue scheduling is even more important, allowing certain programs to be executed which would not be possible otherwise, for example to observe an object each night for a month or a year. Stability of the instrumentation (PFIS[1]—The Prime Focus Imaging Spectrograph—will be always available for the first several years), combined with the queue schedule, will allow one to obtain very uniform data sets over long periods of time, which is very hard to do on other large telescopes.

Here are several GRB projects to consider over the first several years of the SALT operation:

- GRB/SN connection: observe every $z < 0.5$ GRB afterglow (several a year) at 15-20 epochs during the first month after the burst and during 5-10 epochs later. This will answer the questions posed above: what are the GRB progenitors, do they form a uniform class?
- Physics of the afterglow: obtain polarization light curves of several bright afterglows per year for 3-7 days each. This will be important to understand the physics of the GRB emission and to test the jet model of the afterglow.
- Very short timescale variability of GRB afterglows. We have reported short-timescale afterglow variability in two cases (GRB 011211: Holland et al. 2002; GRB 021004: Bersier at al. 2003b). In some cases the afterglows are very smooth (GRB 990510: Stanek et al. 1999; short-timescale variability <0.5% for GRB 020813: Laursen & Stanek 2003). Again, the ability of SALT to perform very rapid photometry (Charles 2004, this volume) is unique among the largest telescopes.

Many other projects can be proposed, here I tried to concentrate on areas where the strengths of SALT would be well utilized. Earlier during the conference, we were encouraged to think about truly large projects ("what would you do with infinite time on SALT") to be done with SALT during the first few years. Being a reasonable person, I decided not to use "infinite time", just 10% of it, to spend one hour *every night* to obtain various kinds of data (depending on the brightness) for any GRB afterglow that is

[1] `http://www.sal.wisc.edu/pfis/`

observable from SALT that night. Such a project would be a major undertaking, but with the expected Swift GRB rate the sample of well observed afterglows would easily *double* in just several months, and there would be many surprises and new exciting results.

ACKNOWLEDGMENTS

I wish to thank my colleagues and friends from the CfA/Notre Dame GRB group, Peter Garnavich, David Bersier, Stephen Holland, Saurabh Jha and Tom Matheson for their excellent collaboration. I also want to thank all the astronomers (about 100 of them so far, too many to be named here) who devoted their telescope time and effort to make this research possible. Some of them collaborated with us on multiple bursts, and I hope we have many future bursts to observe. I thank Tom Matheson, Andy Szentgyorgyi and Joel Hartman for useful comments on this manuscript. I would also like to thank Joanna Mikolajewska and other organizers of the First International Workshop on "Stellar Astrophysics with the World Largest Telescopes" for their invitation and a very enjoyable conference.

REFERENCES

1. Akerlof, C., et al. 1999, Nature, 398, 400
2. Barth, A. J., et al. 2003, ApJ, 584, L47
3. Bersier, D., et al. 2003a, ApJ, 583, L63
4. Bersier, D., et al. 2003b, ApJ, 584, L43
5. Bloom, J. S., et al., 2002, ApJ, 572, L45
6. Bloom, J. S. 2003, in Gamma-Ray Bursts in the Afterglow Era, ed. M. Feroci et al. (astro-ph/0303478)
7. Boella, G., et al. 1997, A&AS, 122, 299
8. Covino, S. et al. 1999, A&A, 348, L1
9. Della Valle, M., et al. 2003, A&A, 406, L33
10. Djorgovski, S. G., et al. 2003, to appear in: Discoveries and Research Prospects from 6-10m Class Telescopes (astro-ph/0301342)
11. Filippenko, A. V., Chornock, R., & Foley, R. J. 2004, GCN Circ. 2517
12. Frail, D. A., et al. 1997, Nature, 389, 261
13. Frail, D. A., et al. 2001, ApJ, 562, L55
14. Galama, T. J., et al. 1998, Nature, 395, 670
15. Garnavich, P. M., et al. 2003a, ApJ, 582, 924
16. Garnavich, P. M., et al. 2003b, IAU Circ. 8108
17. Gehrels, N. et al. 2004, ApJ, 611, 1005
18. Germany, L. M., et al. 2000, ApJ, 533, 320
19. Greiner, J., et al., 2003, GCN Circ. 2020
20. Groot, P. J. et al. 1997, IAU Circ. 6584
21. Hjorth, J., et al., 2003, Nature, 423, 847
22. Holland, S. T., et al. 2002, AJ, 124, 639
23. Jha, S., et al. 2001, ApJ, 554, L155
24. Kawabata, K. S., et al., 2003, ApJl, 593, L19
25. Klebesadel, R. W., Strong, I. B., & Olson, R. A. 1973, ApJ, 182, L85
26. Kosugi, G., et al. 2004, PASJ, 56, 61
27. Lamb, D. Q. 1995, PASP, 107, 1152
28. Laursen, L. T., & Stanek, K. Z. 2003, ApJ, 597, L107
29. Lipkin, Y. M., et al. 2004, ApJ, 606, 381
30. Malesani, D., et al. 2004, ApJl, 609, L5

31. Matheson, T., et al. 2003a, GCN Circ. 2107
32. Matheson, T., et al., 2003b, ApJ, 599, 394
33. Mazzali, P. A., et al. 2003, ApJ, 599, L95
34. Meegan, C. A., et al. 1992, Nature, 355, 143
35. Metzger, M. R., et al., 1997, Nature, 387, 878
36. Mirabal, N., et al. 2003, ApJ, 595, 935
37. Paczyński, B. 1995, PASP, 107, 1167
38. Piran, T. 2004, Reviews of Modern Physics, in press (astro-ph/0405503)
39. Price, P. A., et al. 2003, ApJ, 589, 838
40. Schaefer, B. E., et al. 2003, ApJ, 588, 387
41. Stanek, K. Z., et al., 1999, ApJ, 522, L39
42. Stanek, K. Z., et al., 2003, ApJ, 591, L17
43. van Paradijs, J., et al. 1997, Nature, 386, 686
44. Wijers, R. A. M. J. et al. 1999, ApJ, 523, L33

Testing Stellar Evolution Models of Metal-Poor Stars

Brian Chaboyer

Dartmouth College, Dept. of Physics and Astronomy, 6127 Wilder Lab, 03755 Hanover, NH, USA

Abstract.
I describe how recent observations of metal-poor stars made with large telescopes have lead to new insights into the evolution of metal-poor stars. High dispersion [Fe/H] studies of sub-giant and main sequence turn-off stars in globular clusters have demonstrated that diffusion is inhibited in the envelope of metal-poor stars. This is a puzzling result, as helioseismology provides strong evidence that envelope diffusion occurs in the Sun. Observations of detached double-lined eclipsing binaries allow one to determine the mass and radius of the stars and hence, provide a critical test of stellar evolution models. Observations of detached double-lined eclipsing binaries which are in the main sequence turn-off region of the globular cluster Omega Cen are compared to theoretical stellar evolution models.

CP752, *Stellar Astrophysics with the World's Largest Telescopes: First International Workshop*
edited by J. Mikołajewska and A. Olech

Search for Metal-poor Stars with LAMOST

Gang Zhao

National Astronomical Observatories, Chinese Academy of Sciences
Beijing 100012, CHINA

Abstract. A project of searching for metal-poor stars with Large sky Area Multi-Object fiber Spectroscopic Telescope (LAMOST) is presented. The survey consists of two observational modes for different research targets. One is the search for metal-poor star candidates through low resolution (R = 1000-2000) spectroscopic survey. Another is a follow-up observation with medium resolution (R = 10000-15000) spectrographs aiming on detailed studies of the metal-poor stars.

INTRODUCTION

Metal-poor stars play a very important role in the study of the early Galactic chemical evolution, which also provides the "fossil record" of the creation and evolution of the elements from the earliest times. The most metal-deficient presumably provide information on Pop. III progenitors and such objects must have existed in the past, and perhaps still exist in Galactic halo. In order to progress these kinds of studies, the dedicated survey efforts to expand the number of such stars known, especially below [Fe/H] = –2.0 is very necessary. A sample of stars selected in the survey should cover a wide range of stellar spectral type and temperature. The visual magnitude should be bright enough for easy of high resolution follow-up observation. The numerous samples of the candidate enable us to give the trends and scatter of elemental species. There are two kind of stars are suitable for such survey, namely, dwarfs and giants. The former is a relatively warm and generally un-mixed star, the latter is a relatively cool and generally mixed star. An advantage is these stars are located throughout the Galaxy, which can give us important information of the structure and evolution of the Milky Way.

The systematic discovery of metal-poor stars began with the search for high-velocity objects in catalogs of high proper motion stars. The deliberate search for extremely metal poor stars by objective-prism surveys began with Bond (1970), who recognized the importance of such stars as relics of earliest Galactic times. The method was extended to fainter stars by addition of an interference filter that isolates the Ca II H and K lines with wavelength coverage of 15 nm (Preston & Shectman 1977; Beers 1999). The HK prism survey was finished by using 0.6m Burrell Schmidt telescope in northern hemisphere and 0.6m Curtis Schmidt telescope in southern hemisphere. The survey limited by relatively "bright" magnitude limit with B < 15.5. Beers and his collaborators found about 1000 stars with [Fe/H] < -2.0 by visual detection of Ca II H and K lines. Rhee, Beers & Irwin (1999) re-analyzed the data from objective prism

CP752, *Stellar Astrophysics with the World's Largest Telescopes: First International Workshop*
edited by J. Mikołajewska and A. Olech

spectroscopic survey by the method of combination of automated scans and 2MASS colors. They found about 2000 stars with [Fe/H] < -2.0 through Artificial Neural Network analysis.

Another large scale objective-prism survey is the Hamburg/ESO survey (HES), which started in 1989 with aim of finding bright quasars (B_J > 17.5) by ESO 1m Schmidt telescope. This survey is ideal to search for metal-poor stars since the field with low star number density at high latitude ($|b| > 30°$) . There are total of 383 plates with a 5°×5° deg coverage each. This deep wide-field objective prism survey of the southern sky gives a 7500 square degrees area. With machine scanned and automatically classified method, Christlieb et al. (2001a, b) analyzed totally 4,000,000 stellar spectra. They found about 10000 metal-poor star candidates with [Fe/H] < -2.0 from these spectra since the methods used are highly efficient selection of metal-poor, FHB/A stars, carbon-enhanced stars, etc.

STELLAR ABUNDANCE AND GALACTIC EVOLUTION BASED ON LAMOST SPECTROSCOPIC SURVEY

LAMOST is a national key project started in 1997. Its main features are unique in combining large aperture (4-meter) with wide field of view (5-degree) and equipped unprecedented 4000 fibers spectroscopy. There are two important technical innovations in LAMOST: the active optics for segmented thin mirror and the parallel controllable fiber positioning. The LAMOST can provide 4000 spectra of objects down to V < 20.5 with 1 nm spectral resolution in 1.5 hours exposure, to breakthrough the bottleneck of ever-increasing need for spectroscopic observation in astronomy. An advantage of LAMOST is good for wide field and large sample astronomy. Therefore, we propose a large scale spectroscopic survey project with aim of improvement of our understanding in the filed of the structure and evolution of the Galaxy. One of important scientific goals of this project is the search for metal-poor star candidates with LAMOST low resolution spectrograph and a follow up observation through LAMOST medium resolution spectrograph based on the sample stars selected from the low resolution spectroscopic survey.

Low Resolution Spectroscopic survey with LAMOST

The aim of low resolution spectroscopic survey is to unravel the Galactic structure, the formation history of the Galaxy, the chemical and dynamical evolution of the Galaxy, and dark matter distribution in the Galaxy. An important subject of this survey is the search for metal-poor star candidates based on LAMOST low resolution spectroscopic survey (R = 1000-2000). This extensive amount of data of low resolution spectra will stimulate a revolution in the exploration of Galactic halo. Metallicities of stars may change with the radial direction from the inner halo to the outer halo, if the origin sites of such stars are different. Metallicity histograms of halo stars as a function of radius would allow models of halo formation to be developed and tested.

TABLE1. Characteristic of Low Resolution Spectrographs (LRS).

	Blue Arm		Red Arm	
	R	**W_coverage (nm)**	**R**	**W_coverage (nm)**
full slit	1000	370-590	1000	570-900
1/2 slit	2000	370-590	2000	570-900
1/3 slit	3000	370-590	3000	570-900

There are total 16 LRSs with two 4K × 4K CCD each attached to LAMOST. We give the characteristic of the low resolution spectrographs in Table 1. The consequent columns are the slit width, the resolving power and the wavelength coverage for the blue and red arms respectively. The optical design of LAMOST low resolution spectrograph is given in Figure 1.

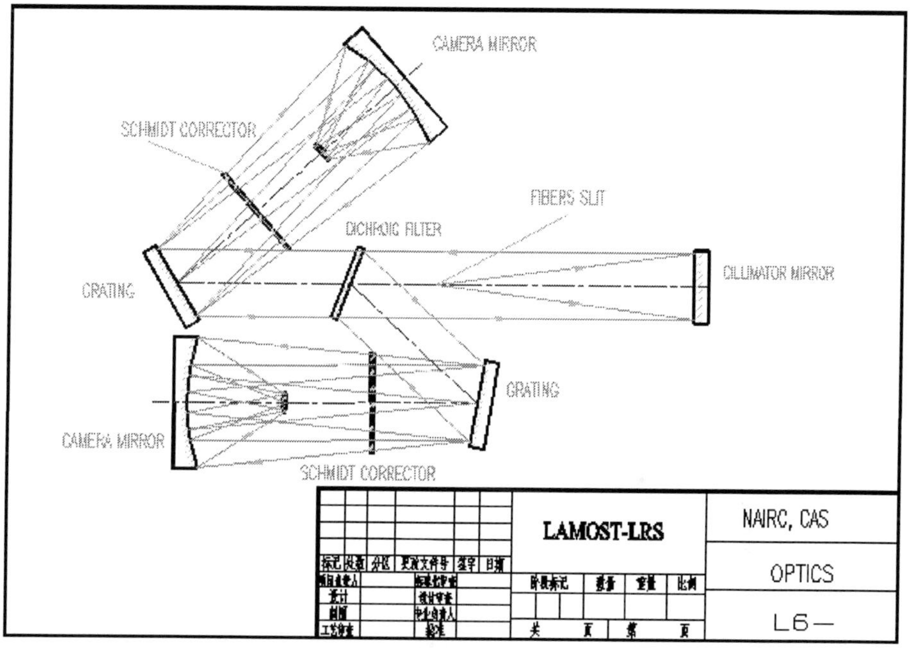

FIGURE 1. The optical design of LAMOST low resolution spectrograph.

We plan to observe around seven millions of stellar objects with 16 low resolution spectrographs within three years survey. Through such survey, various stellar objects can be observed with two resolution modes (R = 1000 and 2000) and a new star catalogue with Sp-type and [Fe/H] can be achieved as well. We expect to find 10^5 metal-poor star candidates with [Fe/H] < -2.0, 10^4 with [Fe/H] < -3.0, 10^3 with [Fe/H] < -4.0, a few tens of stars with [Fe/H] < -5.0 and a few stars with [Fe/H] < -6.0 through the large scale spectroscopic survey.

Medium Resolution Spectroscopic Survey with LAMOST

Medium resolution spectroscopic study for metal-poor stars with R=10000-15000 is a follow up observation based on all high quality metal-poor candidate stars determined by LAMOST low resolution spectroscopic survey. Such medium resolution spectroscopic survey can unravel some important problems, such as light element constraints on Big Bang nucleisynthesis, Alpha- and iron-peak abundances of metal-poor stars, carbon formation in early-generation stars, the direct measurement of elemental yields of Type II SN, identification of astrophysical site(s) of the neutron-capture formation processes, efficiency of mixing processes in the early Galaxy – likely dominance of individual SN II, the hard lower limits on age of Galaxy & Universe through the use heavy-element radioactive chronometers (Th/Eu, U/Eu).

TABLE2. Characteristic of Medium Resolution Spectrographs (MRS).

	Blue Arm		Red Arm	
	R	**W_coverage* (nm)**	**R**	**W_coverage* (nm)**
full slit	7000	510-550	5000	830-890
1/2 slit	14000	510-550	10000	830-890

* other wavelength ranges can be achieved be turn volume phase holographic grating

A MRS consists of a LRS plus a volume phase holographic grating. We give the characteristic of the medium resolution spectrographs in Table 2. The consequent columns in the table are same as in Table 1.

Up till now, there are only 7000 stars with spectral resolving power R > 15000 were observed (one at a time!) over past 20 years. We plan to observe all metal-poor star candidates with B magnitude brighter than 18.5^m determined by LAMOST low resolution spectroscopic survey and by possible SDSS/SEGUE project. We can expect the number of metal-poor stars with relatively high resolution spectra (R ~ 10000 - 15000) will be increased by at least 1-2 magnitudes through such survey project.

ACKNOWLEDGMENTS

This research is supported by the National Natural Science Foundation of China under grant number 10433010 and NKBRSF No. G1999075406.

REFERENCES

1. Beers, T. C., *Ap&SS* **265**, 547 (1999).
2. Bond, H. E., *ApJS* **22**, 117 (1970).
3. Christlieb, N., Wisotzki, L., Reimers, D., Homeier, D., Koester, D. and Heber, U., *A&A* **366**, 898 (2001a)
4. Christlieb, N., Green, P. J., Wisotzki, L.and Reimers, D., *A&A* 366, 898 (2001b).
5. Preston, G., and Shectman, S., *Carnegie Institution of Washington Year Book,* 76 (Washington, DC: Carnegie Inst. Washington), 1977, p149.
6. Rhee, J., Beers, T. C. and Irwin, M. J., *Bulletin of the American Astronomical Society,* **31**, .971 (1999)

LOCAL GROUP GALAXIES

Stellar Populations in the Local Group of Galaxies

Eva K. Grebel

Astronomical Institute of the University of Basel, Venusstrasse 7, CH-4102 Binningen, Switzerland

Abstract. The characteristics and properties of the stellar populations and evolutionary histories of Local Group galaxies are summarized and compared to predictions of cosmological models. No clear signature of the re-ionization epoch is observed; in particular, there is no cessation of star formation activity in low-mass dwarf galaxies at the end of re-ionization. Arguments against the morphological transformation of dwarf irregular into dwarf spheroidal galaxies are derived from their pronounced evolutionary differences at early epochs as evidenced by the offset in the metallicity-luminosity relation between gas-rich and gas-poor dwarfs. While there is increasing evidence for past and ongoing accretion events the overall importance of dwarf galaxies as building blocks remains unclear considering their differences in modes of star formation and detailed chemistry.

THE LOCAL GROUP

The Local Group of galaxies is dominated by two massive spirals, M31 and the Milky Way. The Local Group (LG) is our immediate cosmic neighborhood, a sparse galaxy group with a zero-velocity radius of 1 Mpc (Karachentsev et al. 2002a). Two dominant galaxies also characterize a number of other nearby galaxy groups, and zero-velocity radii of 1 to 1.2 Mpc are typical for these poor groups. Thirty-eight galaxies are currently known to be located within the LG's radius, but the galaxy census of the LG is still incomplete. While all of its luminous members have been detected, new faint member candidates with very low surface brightnesses are still being discovered (e.g., Zucker et al. 2004a). In spite of its small size and its small number of galaxies the LG plays a prominent role in astrophysical research. For reviews of the LG, see Grebel (1997, 1999, 2000, 2001), Mateo (1998), and van den Bergh (1999, 2000). In the present contribution, I will mainly concentrate on the dwarf galaxies of the LG.

The proximity of the galaxies of the LG permits us to study them at the highest possible resolution, enabling us to resolve these galaxies into individual stars down to very faint magnitudes. For example, with the Hubble Space Telescope, point sources down to an apparent magnitude of ≈ 30 have been detected in M31 (Brown et al. 2003), corresponding to the magnitudes of stars well below the Population II main-sequence turn-off at that distance. To date, the LG is the only location where we can observe the lowest stellar masses, measure the oldest stellar ages, determine metallicities and element abundance ratios of individual stars of high ages, and measure detailed stellar kinematics. In other words, the LG is unique in affording the highest level of detail and accuracy for stellar population studies. Much of this research has only become possible in recent years thanks to the superior angular resolution and sensitivity of the Hubble Space Telescope (HST), and thanks to the advent of powerful ground-based

CP752, *Stellar Astrophysics with the World's Largest Telescopes: First International Workshop*
edited by J. Mikołajewska and A. Olech

optical/near infrared 8 to 10m class telescopes, which provide sufficient sensitivity for medium and high-resolution stellar spectroscopy of individual stars in nearby galaxies. It is now becoming possible to uncover the evolutionary histories of these galaxies in unprecedented detail.

Furthermore, the LG contains a variety of different galaxy types. While massive early-type galaxies are absent, the LG still provides us with galaxies covering five orders of magnitude in galaxy masses, a variety of different morphological types, very different star formation histories, and a range of different dominant ages and metallicities. Within the small volume of the LG, galaxies occur in environments ranging from immediate proximity to massive galaxies to fairly isolated locations without nearby neighbors. Hence we can expect to eventually obtain first-hand measurements of the impact of environment on galaxy evolution, at least on small scales.

All in all, by studying the stellar populations of galaxies in the LG and by combining them with detailed knowledge of the other components of these galaxies such as gas, dust, and non-luminous matter we can learn about galaxy properties at the highest possible level of detail. The LG permits us to study stellar evolution at a range of different metallicities and ages not afforded in that combination in the Milky Way; e.g., the evolution of young, massive stars at low metallicities. Moreover, the LG permits us to carry out tests of cosmological galaxy evolution theories, for instance with respect to time scales and number and magnitude of accretion events. Hence here we can conduct "near-field cosmology" (see Freeman & Bland-Hawthorn 2002). The LG is particularly well-suited to learn more about the formation and evolution of disk galaxies, whose theoretical foundations are still poorly understood. Understanding all of these processes in an environment where we can learn about them in detail is the precondition for understanding distant, unresolved galaxies.

The galaxy content of the Local Group

We currently know of 38 galaxies within the zero-velocity surface of the Local Group. Whether the outermost galaxies are indeed gravitationally bound to the LG remains unclear. Indeed, reliable orbits are not yet known for any of the LG galaxies, and proper motion measurements exist only for the closest Milky Way companions. The uncertainties of these measurements are still large, but this situation should change once data from the planned astrometric satellite missions – Gaia (an ESA mission that will scan repeatedly the entire sky) and the Space Interferometry Mission (SIM, a NASA mission that will perform pointed observations with even higher accuracy) – become available. These missions will probably not start before 2011.

As mentioned before, the LG galaxy census is still incomplete. Apart from missing orbital information, galaxies of very faint surface brightness are still being detected as sky surveys become more sensitive or as data mining techniques are being improved. During the past six years, this led to the discovery of one fairly isolated new dwarf spheroidal (dSph) galaxy in the Local Group (Cetus; Whiting, Irwin, & Hau 1999), and to the discovery of four new dSph companions of M31 (Armandroff, Davies, & Jacoby 1998; Armandroff, Jacoby, & Davies 1999; Karachentsev & Karachentseva 1999, Grebel

& Guhathakurta 1999, Zucker et al. 2004a, Harbeck et al. 2005). The conclusion of data mining existing photographic all-sky surveys (Karachentsev et al. 2000; Whiting, Hau, & Irwin 2002) makes the existence of additional faint, Draco- or Ursa-Minor-like dSphs seem unlikely unless they are hidden behind regions of high Galactic extinction, particularly in the zone of avoidance. On the other hand, the discovery of And IX (Zucker et al. 2004a) with an unusually low surface brightness of only $\mu_{V,0} \sim 26.8$ mag arcsec^{-2} indicates that additional very sparse, very faint dwarfs may yet have to be discovered. At present, though, there is little reason to believe that there may still be hundreds of yet to be detected faint objects present in the LG.

Very faint, sparse, low-surface-brightness objects like And IX, which appear to be dominated by ancient, metal-poor stars (Harbeck et al. 2005) are of immense interest from the point of view of galaxy evolution. Do they show signatures of cosmic re-ionization squelching? Do they contain the purest, most metal-poor, most ancient populations known? Or do they show evidence for population gradients in their old population as were found in a number of other ancient dSphs (Harbeck et al. 2001)? Do these objects show large metallicity spreads as measured in other seemingly purely old dSphs (Shetrone, Côte, & Sargent 2001), which in turn indicate extended episodes of star formation (Ikuta & Arimoto 2002)? How did these objects survive? Are they entirely dominated by dark matter? Are they regular dSph galaxies that simply are fainter and sparser than the previously known members of this galaxy class? Yet from a cosmological point of view, objects of this type may appear to be of little interest since they may not significantly contribute toward solving the "missing satellite problem" (Klypin et al. 1999; Moore et al. 1999) unless detected in vast numbers.

Not only low surface brightness, scarcity of stars, high foreground extinction, and unknown orbits for distant LG member candidates contribute to an incomplete LG galaxy census, but also accretion events. We still know little about the actual number of accretion events, about the times when these events occurred, and about the nature of the accreted galaxies (see also the Section 1.4 and the contribution of Majewski in these proceedings). The number of 38 probable LG members comprises all galaxies listed in Grebel, Gallagher, & Harbeck (2003) including the Sagittarius dwarf galaxy, which is currently being accreted by the Milky Way. In addition, it includes the giant stream around M31 (Ibata et al. 2001), which is very likely the remnant of a disrupted galaxy, and the newly discovered M31 satellite And IX (Zucker et al. 2004a). *Not* included are other possible streams and accretion remnants such as the Monoceros feature (Newberg et al. 2002) and Canis Major (Martin et al. 2004a) since their nature as dwarf galaxy remnants remains disputed (see Section 1.4). Depending on the nature of these and other stellar overdensities and kinematically identified features in the Milky Way and in M31, the LG census may need to be increased by at least up to six or more galaxies if one wishes to account for still measurable accretion events. If these features are confirmed as minor mergers, then it becomes a matter of definition whether their progenitors should be included in the current LG member census.

Galaxy types in the Local Group

Approximately 90% of the LG's luminosity is contributed by its three spiral galaxies, but the vast majority of the LG galaxy population are satellite and dwarf galaxies. The boundaries between larger galaxies and dwarf galaxies are poorly defined. We adopt here the usual convention of calling every galaxy with an absolute magnitude of $M_V > -18$ mag a dwarf galaxy. This simple magnitude criterion then leaves us with 34 dwarf galaxies (though admittedly the absolute magnitude of the progenitor of M31's "great stream" is yet unknown, and that of the Sagittarius dSph is poorly constrained). Apart from the three spirals, only the Large Magellanic Cloud is more luminous than the above luminosity threshold. Yet M32 is often not considered a dwarf galaxy despite its lower luminosity since it exhibits similar structural properties as giant ellipticals.

Most of the LG galaxies belong to one of four basic classes: Spirals, dwarf irregular galaxies, dwarf elliptical galaxies, and dSphs. In the following paragraphs the main characteristics of the dwarf galaxy classes are summarized following Grebel (2001):

Dwarf irregular galaxies (dIrrs) are gas-rich galaxies with an irregular optical appearance usually dominated by scattered H II regions. They typically have V-band surface brightnesses of $\mu_V \lesssim 23$ mag arcsec^{-2}, H I masses of $M_{\rm HI} \lesssim 10^9$ M$_\odot$, and total masses of $M_{\rm tot} \lesssim 10^{10}$ M$_\odot$. The stellar populations of dIrrs range from ancient stars with ages > 10 Gyr to ongoing star formation. Star formation appears to have proceeded largely continuously, although amplitude variations in the intensity or rate of star formation may reach a factor of three. The H I distribution is usually clumpy and more extended than even the oldest stellar populations. When considering only the distribution of older stellar populations (i.e., of red giants and red clump stars, which trace populations older than ~ 1 Gyr), irregulars and dIrrs exhibit a highly regular distribution shaped like an (elongated) disk (LMC; van der Marel 2001) or like a spheroid (example: Small Magellanic Cloud (SMC); Zaritsky et al. 2000). In low-mass dIrrs gas and stars may exhibit distinct spatial distributions and different kinematic properties. Metallicities tend to increase with decreasing age in the more massive dIrrs, indicative of enrichment and of an age-metallicity relation (see also Fig. 3 in Grebel 2004). For low-mass dIrrs detailed measurements are still lacking. Solid body rotation is common among the more massive dIrrs, whereas low-mass dIrrs seem to be dominated by random motions without evidence of rotation. DIrrs are found in galaxy clusters, groups and in the field.

Dwarf elliptical galaxies (dEs) are spherical or elliptical in appearance, tend to be found near massive galaxies (in the Local Group they are all companions of M31), usually have little or no detectable gas, and tend not to be rotationally supported. Note that examples of rotating dEs beyond the Local Group have been found (e.g., Pedraz et al. 2002). DEs are compact galaxies with high central stellar densities. They are typically fainter than $M_V = -17$ mag, have $\mu_V \lesssim 21$ mag arcsec^{-2}, $M_{\rm HI} \lesssim 10^8$ M$_\odot$, and $M_{tot} \lesssim 10^9$ M$_\odot$. DEs may contain conspicuous nuclei (nucleated dEs or dE,N) that may contribute up to 20% of the galaxy's light. The fraction of dE,N is higher among the more luminous dEs. An example for a non-nucleated dE in the Local Group is NGC 185, whereas NGC 205 is a dE,N. Sérsic's (1968) generalization of a de Vaucouleurs $r^{1/4}$ law and exponential profiles describe the surface density profiles of nucleated and non-nucleated dEs and dSphs best (Jerjen et al. 2000). DEs typically contain old and intermediate-age populations (i.e., populations older than 10 Gyr and populations in an age range of ~ 2 to 10 Gyr), but the fractions of these populations vary, and even present-day star formation may be observed.

Dwarf spheroidal galaxies (dSphs) are diffuse, gas-deficient, low-surface-brightness dwarfs with very little central concentration. They are not always distinguished from dEs in the literature. DSphs are characterized by $M_V \gtrsim -14$ mag, $\mu_V \gtrsim 22$ mag arcsec^{-2}, $M_{\rm HI} \lesssim 10^5$ M$_\odot$, and $M_{tot} \sim 10^7$ M$_\odot$. They include the optically faintest galaxies known. Their stellar populations tend to be either almost purely old or a mix of old and intermediate-age populations. The luminosity functions (and by inference the mass functions) of dSphs have been found to be "normal" and in excellent agreement with those of Galactic globular clusters (Grillmair et al. 1998; Wyse et al. 2002). DSphs are usually found in close proximity of massive galaxies (counterexamples in the LG: Cetus, Tucana) and are generally not supported by rotation. Since most published measurements concentrated on the central regions of dSphs, the possibility of slowly rotating dSphs cannot be excluded. The central velocity dispersions of dSphs indicate the presence of a significant dark component when virial equilibrium is assumed. However, not all dSphs are in virial equilibrium (e.g., Ursa Minor shows indications of being tidally distorted by the Milky Way; for instance, Palma et al. 2003). Interestingly, the radial velocity dispersion profiles of dSphs show a marked drop at large radii (Wilkinson et al. 2004). Its interpretation and the apparent existence of a kinematically cold stellar population at the outermost radii is not yet understood. The metallicity–luminosity relations of dSphs and dIrrs show the usual trend of increasing metallicity with increasing galaxy luminosity, but the relations are offset from each other: DSphs have higher mean stellar metallicities at a given optical luminosity (Grebel et al. 2003 and Section 3.1).

Morphological segregation

As already indicated, the distribution of the different types of galaxies within the LG is not random, but is determined by the location of the two most massive spiral galaxies in the LG, M31 and the Milky Way. A sketch of the three-dimensional distribution of galaxies within the LG is shown in Grebel (1999; in Fig. 3). In Fig. 1 the distribution of gas-poor, low-mass dwarfs and of gas-rich, higher-mass dwarfs is shown. The gas-poor dwarfs – primarily dSphs – are strongly clustered around the closest massive spiral, whereas the gas-rich dwarfs (dIrrs) exhibit less of a tendency toward close concentration around massive galaxies and are more widely distributed. The differing distribution of different morphological galaxy types is also known as morphological segregation and is observed in nearby galaxy groups and galaxy clusters as well.

Similarly, the H I mass of dwarf galaxies tends to increase with increasing distance to a massive primary (Grebel et al. 2003; their Fig. 3). These trends indicate that environment may have a significant impact on the evolution of low-mass galaxies. Indeed, it is tempting to speculate that these trends are the result of morphological transformations due to the influence of massive galaxies, e.g., via tidal or ram pressure stripping (see Mayer et al. 2001 for simulations). One vital ingredient to verify this hypothesis is once again information about the orbits of the companion galaxies.

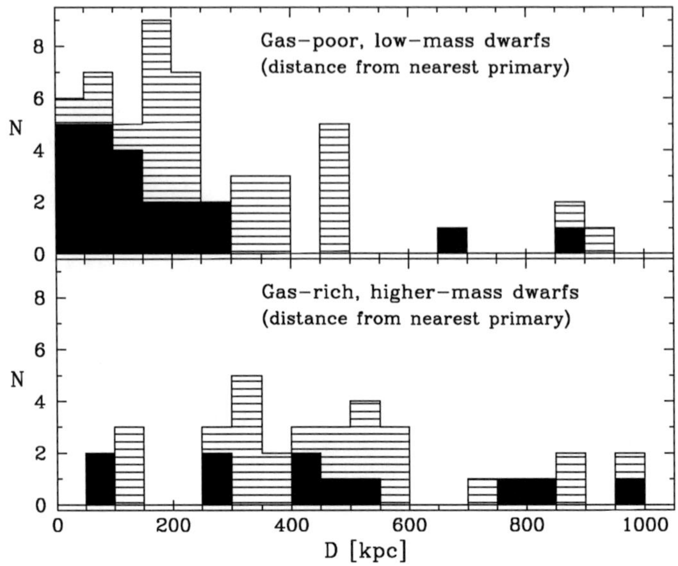

FIGURE 1. Morphological segregation in the number distribution N of different types of galaxies in the Local Group (solid histograms) and in the M81 and Centaurus A groups (hashed histograms) as a function of distance D to the closest massive primary (updated version of Fig. 1 in Grebel 2004).

Direct evidence for harassment and accretion

If accretion is the primary mechanism for the growth and evolution of massive galaxies as predicted by cosmological models, then we should be able to find evidence for these processes in our immediate neighborhood. (1) The study of the structural properties of nearby galaxies can reveal whether external tidal forces are distorting them. (2) The detection of extratidal stars and streams around and within massive galaxies is evidence for ongoing harassment and accretion events. (3) The stellar content, population properties, and chemistry of nearby galaxies allow us to constrain to what extent these kinds of objects could have contributed as building blocks to more massive galaxies.

The clearest evidence for ongoing accretion are the extended tidal stream of the Sagittarius dSph galaxy (Ibata, Gilmore, & Irwin 1994), which has now been traced around the entire Milky Way using M giants identified in the Two Micron All-Sky Survey (2MASS) (Majewski et al. 2003), and the giant stream of metal-rich giants in the halo of M31 (Ibata et al. 2001; Ferguson et al. 2002). Additional stellar overdensities have been detected in the Milky Way using various photometric data sets including 2MASS and the Sloan Digital Sky Survey (SDSS): The Monoceros feature (Newberg et al. 2002; Yanny et al. 2003), which may be a tidal tail connected with the Canis Major overdensity (Martin et al. 2004a). The interpretation of Canis Major is disputed; suggestions include that it is part of the Galactic warp or flare (Momany et al. 2004) or indeed the center of another possibly disrupted dSph within the Milky Way (Martin et al. 2004a, 2004b). Additional Galactic stellar overdensities have been identified (for instance, Triangulum-

Andromeda; Rocha-Pinto et al. 2004, which may be part of the tidal tail of a more distant disrupted dwarf). Other suggestions of evidence of dwarf galaxy accretion are based on the identification of moving groups and radial velocity surveys (e.g., Gilmore, Wyse, & Norris 2002). However, not only disrupted dwarf galaxies, but even globular clusters in advanced stages of accretion may produce extended stellar tidal tails (Odenkirchen et al. 2001; 2003). These tidal features are valuable also as tracers of the Galactic potential. The most luminous and most massive Galactic globular cluster, ω Centauri, contains a range of different ages and a large metallicity spread. A popular explanation for its unusual properties is that ω Cen may be the core of an accreted dwarf galaxy (see van Leeuwen, Hughes, & Piotto 2002 for details). — Possible additional tidal features in and near M31 have been reported by Morrison et al. (2003) and Zucker et al. (2004b).

The Magellanic Clouds and the Milky Way are interacting with each other as evidenced by, e.g., the gaseous Magellanic Stream and the Magellanic Bridge (e.g., Brüns et al. 2005), although the interpretation of these features as being primarily due to tidal (e.g., Putman et al. 1998) or ram pressure effects (Mastropietro et al. 2005) remains controversial. The twisted isophotes of the M31 companions M32 and NGC 205 may be caused by tidal interaction with M31 (Choir, Guhathakurta, & Johnston 2002). The nearby Galactic dSph satellite Ursa Minor shows a distorted, S-shaped surface density profile possibly caused by tidal interaction with the Milky Way (Palma et al. 2003). On the other hand, the drop in the velocity dispersion profiles of Draco and Ursa Minor at large radii (Wilkinson et al. 2004) and the lack of a large depth extent of Draco (Klessen, Grebel, & Harbeck 2003) would seem to argue against ongoing tidal *disruption*.

THE EARLIEST EPOCH OF STAR FORMATION

According to hierarchical structure formation scenarios, low-mass systems should have been the first sites of star formation in the Universe. The first star formation events may have occurred as early as at a redshift of $z \sim 30$, an epoch unobservable for us with our present tools. Larger systems should then have formed through hierarchical merging of smaller systems, leading to the idea of dwarf galaxies as building blocks of more massive galaxies. Hence one important test to carry out is to compare the properties of the old populations in dwarf galaxies with those in massive galaxies to investigate how similar or dissimilar they are. If today's dwarf galaxies are the few survivors of a once much more numerous, since accreted low-mass galaxy population, and if dwarf galaxy accretion is the primary process governing the formation of more massive galaxies, then detailed studies of their old stellar populations should reveal very similar properties.

A number of cosmological models predict that cosmic re-ionization will squelch star formation in low-mass substructures, and that galaxies less massive than 10^8 to 10^9 $M_\odot$ will lose their star-forming material through photoevaporation during re-ionization (e.g., Ferraro & Tolstoy 2000; Dekel & Woo 2003; Susa & Umemura 2004). As a consequence, low-mass galaxies must form their stars prior to re-ionization (e.g., Tassis et al. 2003) and need to contain ancient populations. Furthermore, one would expect a sharp drop and indeed a complete cessation of star-forming activity after re-ionization is complete. A third consequence is that the oldest populations in high-mass galaxies must be either as old as those in low-mass galaxies, or younger.

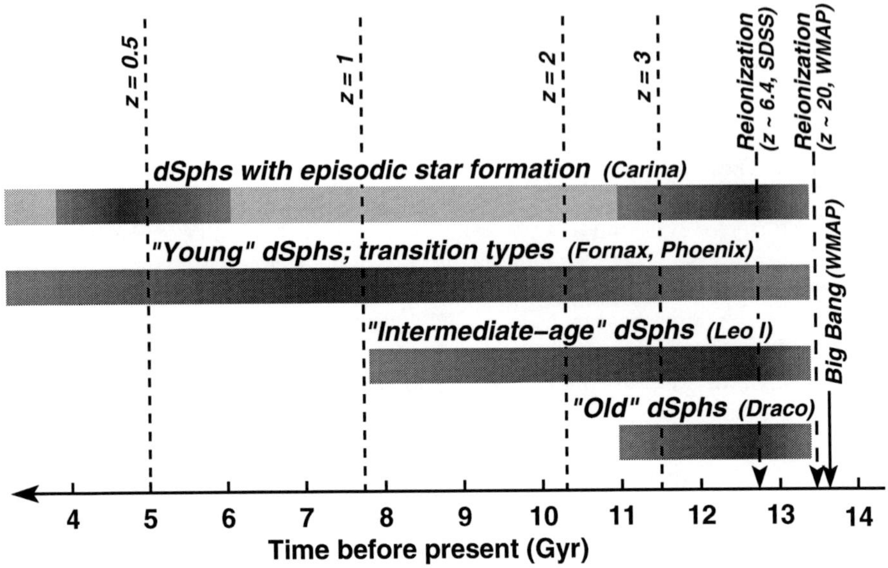

FIGURE 2. Bar diagram indicating the approximate duration of star formation episodes in low-mass galaxies ($\sim 10^7$ $M_\odot$). The approximate beginning and end of the re-ionization epoch are indicated (based on results from WMAP and from the Sloan Digital Sky Survey). The predicted cessation of star formation in low-mass galaxies, esp. in dSphs, is not observed. For more details, see also Grebel & Gallagher (2004).

These are testable predictions that can be investigated by exploiting the fossil stellar record in nearby galaxies. The LG is an ideal target since here the oldest populations are resolved and can be accessed with HST and increasingly also with large ground-based telescopes operating with high angular resolution. This requires age dating of old populations. The most accurate ages can be obtained for resolved stellar populations. For old populations the most age-sensitive feature is the old main-sequence turn-off (MSTO), where via differential age-dating techniques internal accuracies of less than 1 Gyr can be obtained. Comparison objects are usually ancient Galactic globular clusters of the same metallicity as the target population. Absolute ages are more difficult to determine since here one needs to rely on isochrones, which makes the resulting ages model-dependent. But also differential techniques have a number of drawbacks: They require very deep, high-quality photometry reaching at least 2 mag below the MSTO, which is a challenge in more distant galaxies. They require stellar populations sufficiently numerous to produce a measurable MSTO, which necessarily limits us to Population II stars (note that to date not a single Population III star candidate has ever been detected beyond the Milky Way). Ideally, one wishes to compare populations with the same [α/Fe] ratio. (Establishing to what extent this condition is met and what the consequences are will be an important area for large ground-based telescopes in the coming years.) Other effects such as diffusion may further affect relative measurements; see Chaboyer's contribution in these proceedings for details.

For a more detailed discussion of the results of differential age dating as applied to ancient Milky Way populations and nearby dwarfs I refer to Grebel (2000) and Grebel & Gallagher (2004). Here only the results shall be summarized, which are based on Galactic and extragalactic globular clusters and field populations with MSTO photometry: **(1) Old populations are ubiquitous, but their fractions vary. There is not a single dwarf galaxy studied in sufficient detail *without* an old population. (2) There is evidence for a common epoch of star formation. The ancient Population II in the Milky Way, the LMC, and the Galactic dSphs are coeval within $\sim$ 1 Gyr** (which is consistent with the building block scenario). **(3) In contrast to predictions from cold dark matter models, no cessation of star formation activity is observed during or after re-ionization. (4) Instead, even the least massive galaxies known, the dSphs, show evidence for star formation extending over many Gyr.** This holds even for those dSphs entirely dominated by very old populations – their metallicity spread requires enrichment due to star formation extending over several Gyr.

It is certainly correct to point out that the accuracy of age determinations does not permit us yet to confidently state when exactly early star formation began and ended in dSphs. *Nevertheless, the presence of ancient populations in dSphs is a fact.* The *large metallicity spread* in old dSphs has been proven spectroscopically. Furthermore, *the presence of intermediate-age populations in addition to ancient stars in many though not all dSphs cannot be refuted.* In only one of these dSphs (Carina) clearly episodic star formation with well-separated "bursts" has been observed (Smecker-Hane et al. 1994); in all others star formation appears to have proceeded fairly continuously. The characteristics of the star formation and enrichment histories of dSphs contradict the cosmological predictions mentioned earlier, particularly the suggestion of complete photoevaporation of baryonic material not yet turned into stars.

There are several possible ways out of this quandary. Either the above quoted cosmological models are wrong, or do not take other effects that might prevent photoevaporation into account, or the galaxies we observe today as dSphs once were substantially more massive. In case if the latter, today's dSphs would once have needed to be at least a factor of 100 more massive. An oversimplified estimate shows that at the times when dSphs were forming stars, they should have had at least 10 times more baryonic mass if we assume a star formation efficiency of 10%. Their present-day baryonic masses are of the order of 10^6 $M_\odot$, so an increase by one order of magnitude would not yet change the overall estimated masses of dSphs drastically, but it would considerably alter their mass-to-light ratios. More detailed calculations and more observational data are needed to investigate whether substantially more massive dSph progenitors are plausible. The differences in the metallicity-luminosity relation of dSphs and dIrrs seem to rule out dIrrs as progenitors of dSphs (Grebel et al. 2003).

STAR FORMATION HISTORIES AND METALLICITIES

Much has been said already about star formation histories in the preceding sections. In the past years, considerable effort has been spent on uncovering star formation histories of nearby dwarf galaxies by photometric means, i.e., via color-magnitude diagrams of the resolved stellar populations of the stars in these galaxies. Basically, very detailed star formation histories are derived photometrically through synthetic color-magnitude diagrams (CMDs) generated from isochrones assuming an initial mass function and a variety of different time-dependent star formation and chemical evolution histories. The synthetic CMDs are compared to the observed CMDs using various statistical techniques until the best match is found (e.g., Tosi et al. 1991; Gallart et al. 1996; Tolstoy & Saha 1996; Dolphin 1997; Holtzman et al. 1997). These methods necessarily have to make a number of assumptions (such as which initial mass function and which binary fraction to adopt), rely strongly on the chosen set of isochrones, and are plagued by the age-metallicity degeneracy, but yield surprisingly similar results in spite of the different approaches of different groups (see contributions in Lejeune & Fernandes 2002).

The information gained from deep CMDs can be complemented by information provided by individual age tracer stars such as Wolf-Rayet stars, carbon stars, RR Lyrae stars, etc. (see Grebel 1997, 1999) to yield a more complete picture of the star formation histories of the galaxies in question. In particular, such tracer populations are valuable when their specific host populations are too sparse to appear prominently in the CMDs. Furthermore, star formation histories vary as a function of position with the oldest populations being typically the most extended ones (e.g., Harbeck et al. 2001), hence ideally one wishes to cover the entire target galaxy.

Our current knowledge about the modes of star formation in LG dwarf galaxies can be summarized as follows: Irregular and dIrr galaxies are characterized by largely continuous star formation with amplitude variations of factors 2–3, governed mainly by internal, local processes. In the more massive irregulars with large gas reservoirs, star formation is likely to proceed for another Hubble time (Hunter 1997). Low-mass so-called dIrr/dSph transition-type galaxies are characterized by currently very low star formation rates and may eventually turn into quiescent dSphs. A detailed review of the evolutionary histories of these galaxies, large-scale and small-scale star formation properties, and their chemical evolution is given in Grebel (2004).

DE and dSph galaxies also tend to have continuous star formation rates, but with decreasing intensity. Some had their peak activity at very early ages, others several Gyr ago. Star formation tends to be longest-lasting in the centers of these galaxies, and in a number of cases age (and possibly metallicity) gradients are observed (Harbeck et al. 2001). Repetitive or episodic star formation, as mentioned earlier, has so far only found in Carina. For more detailed reviews, see Grebel (2000, 2001). A major puzzle is the surprising lack of gas in dSph galaxies, where (with the exception of Sculptor; Bouchard, Carignan, & Mashchenko 2003) only upper limits for neutral and ionized gas could be determined (Gallagher et al. 2003; Grebel et al. 2003, and references therein). Interestingly, the limits for neutral hydrogen lie well below the amounts expected from gas loss from the old red giants in the dSphs. Fornax, the only dSph galaxy with star formation as recent as $\sim$ 200 Myr ago (Grebel & Stetson 1999), surprisingly also appears to be devoid of gas. **It is neither understood what caused the gas loss to begin with nor how it is sustained**, especially in the distant, isolated dSphs like Cetus and Tucana.

When illustrating star formation histories via population boxes (e.g., Grebel 1997, 1999, 2000), it becomes quickly obvious that no two dwarf galaxies – not even low-mass dSphs – share the same star formation history. Each galaxy needs to be considered as an individual with regard to the time scales of its star formation and the degree of enrichment. A trend of increasing intermediate-age population fractions with increasing distance from the Milky Way among dSphs was first noted by van den Bergh (1994), who attributed it to the possible environmental impact of the Milky Way. Star-forming material might have been removed earlier on from the closer Galactic companions via ram-pressure or tidal stripping, supernova-driven winds or the high UV-flux from the proto-Milky Way. On the other hand, if environment is primarily responsible for gas-poor dSphs, then the existence of the isolated dSphs Cetus and Tucana is difficult to understand. Again, knowledge of the orbits of these galaxies would be very helpful.

We may then turn to the M31 dSph companions, which cover a similar range of distances as the Galactic ones. **Interestingly, these dSphs do *not* show any clear correlation between star formation history and present-day distance to M31**. Their lack of a pronounced red clump and of substantial number carbon stars shows a lack of any prominent intermediate-age population regardless of their distance (Harbeck et al. 2001, 2004, 2005). Going one step further, the comparison of the stellar populations in M31's dSphs and of M31's halo shows that the dSphs cannot have been primary building blocks of M31's halo since it was found to be dominated by intermediate-age, comparatively metal-rich populations (Brown et al. 2003). (An old, metal-poor halo population, however, is present as well; Brown et al. 2004).

Mean metallicities and the metallicity-luminosity relation

Galaxies generally obey a metallicity-luminosity relation such that more luminous (and potentially more massive) galaxies are more metal-rich than faint galaxies with presumably more shallow potential wells. DIrrs and dSphs differ in their metallicity-luminosity relations (e.g., Binggeli 1994). Originally this was discovered when comparing nebular H II region oxygen abundances of dIrrs with stellar metallicity estimates of gas-poor dwarfs such as dSphs. However, in making this comparison, very different populations are compared, and very different metallicity tracers are used. Nebular oxygen abundances cannot easily be translated into mean stellar metallicities and vice versa. Mean stellar metallicities usually rely on the measurement of iron or of an element that has been found to be an excellent tracer of iron in certain populations, such as the near-infrared Ca II triplet (Armandroff & Da Costa 1991; Rutledge, Hesser, & Stetson 1997; Cole et al. 2004). These mean stellar metallicities are usually quoted as [Fe/H] or [Me/H] (to indicate that primarily *metallicity* as opposed to solely *Fe* is meant).

In order to avoid the uncertain O to Fe conversion, Richer, McCall, & Stasinska (1998) published a metallicity-luminosity relation based entirely on nebular O measurements: H II region abundances in dIrrs and planetary nebula (PN) abundances in dEs and dSphs. The offset between the resulting relations remained; however, one still compares different populations with each other. H II region abundances trace the present-day abundances of the most recently formed population of stars. PNe trace primarily intermediate-age populations with ages of at least several 100 Myr if not several Gyr. Furthermore, PNe have only been detected in two dSphs to date (Fornax and Sagittarius). Obviously one cannot expect to measure H II regions in the gas-deficient dSphs.

In order to compare not only *mean stellar metallicities* in dIrrs and dSphs, but also the metallicities of the *same populations* (i.e., of stars of similar age), we instead concentrated on old Population II giants, which as mentioned earlier have been detected in all LG dwarf galaxies. We used (1) *old red giants* in dSphs and in the outskirts of dIrrs (where old populations dominate), (2) *spectroscopic abundances* wherever available (from own Keck LRIS measurements and literature data from studies conducted at ESO, NOAO, and Keck), and (3) *photometric abundances* elsewhere (from comparison with globular cluster fiducials applied to our own deep HST data from a WFPC2 snapshot survey and archival or literature data). While the degree of homogeneity of the resulting data set is not ideal, it is the best currently available one and entirely based on well-calibrated empirical indicators.

The resulting metallicity-luminosity relationship shows that even when confined to old populations, there is a considerable offset between dSphs and dIrrs (Grebel et al. 2003). **At the same galaxy luminosity, the old populations of dSphs are more metal-rich than those of dIrrs**. This indicates that in contrast to dIrrs, dSphs must have experienced fairly rapid early enrichment. Together with various other factors, **these evolutionary differences make normal dIrrs unlikely progenitors of dSphs** (see also Binggeli 1994). DIrr/dSph transition-type galaxies, on the other hand, seem fairly plausible progenitors as explained in more detail in Grebel et al. (2003).

Detailed abundance ratios

Mean spectroscopic metallicities of individual stars provide crucial constraints on otherwise photometrically derived star formation histories and allow one to break the age-metallicity degeneracy – one of the prime areas of study with the new large telescopes.

Another very important area for large telescopes is high-resolution spectroscopy to measure individual abundance ratios. In particular, the determination of Fe, α-, r-, and s-process element abundance ratios makes it possible to measure the modes and rates of star formation spectroscopically: the relative contribution of supernovae of Type II vs. Ia, and that of AGB stars, at different times during the evolution of the target galaxy. Importantly, this research permits one to compare the abundance ratios of different types of dwarf galaxies with those measured in various Galactic components and to constrain the building block scenario from the chemical point of view.

The [α/Fe] ratios in dSphs (and dIrrs) at a given [Fe/H] are lower than those measured in the Galactic halo, indicating either low star formation rates in the dwarfs, loss of metals, or a larger contribution from supernovae of Type Ia. This is **strong evidence against present-day dSphs as the dominant contributors to the build-up of the Galactic halo** (Shetrone et al. 2001). For a discussion of the implications of recent results for nucleosynthesis and galaxy evolution, see also Venn et al. (2004). Extending these kinds of measurements and adding kinematic data as well is rapidly becoming one of the major research areas for the world's largest telescopes such as SALT, nicely complemented by ongoing and future space missions such as HST, JWST, and Gaia.

ACKNOWLEDGMENTS

I would like to thank Joanna Mikolajewska for organizing a very stimulating conference and for her patience while this contribution was finished.

REFERENCES

1. T. E. Armandroff, and G. S. Da Costa, *AJ*, **101**, 1329–1337
2. T. E. Armandroff, J. E. Davies, and G. H. Jacoby, *AJ*, **116**, 2287–2296 (1998).
3. T. E. Armandroff, G. H. Jacoby, and J. E. Davies, *AJ*, **118**, 1220–1229 (1999).
4. B. Binggeli, "Dwarf galaxies: a morphological overview", in *Panchromatic View of Galaxies. Their Evolutionary Puzzle*, edited by G. Hensler, C. Theis, & J. S. Gallagher, Editions Frontiers, Gif-sur-Yvette, 1994, pp. 173–191
5. A. Bouchard, C. Carignan, and S. Mashchenko, *AJ*, **126**, 1295–1304 (2003).
6. T. M. Brown, H. C. Ferguson, E. Smith, R. A. Kimble, A. V. Sweigart, A. Renzini, R. M. Rich, and D. A. VandenBerg, *ApJ*, **592**, L17–L20 (2003).
7. T. M. Brown, H. C. Ferguson, E. Smith, R. A. Kimble, A. V. Sweigart, A. Renzini, and R. M. Rich, *AJ*, **127**, 2738–2752 (2004).
8. C. Brüns, et al., *A&A*, in press (astro-ph/0411453), (2005).
9. P. I. Choi, P. Guhathakurta, and K. V. Johnston, *AJ*, **124**, 310–331 (2002).
10. A. A. Cole, T. A. Smecker-Hane, E. Tolstoy, T. L. Bosler, and J. S. Gallagher, *MNRAS*, **347**, 367–379 (2004).
11. A. Dekel, and J. Woo, *MNRAS*, **344**, 1131–1144 (2003).
12. A. E. Dolphin, *NewA*, **2**, 397–409 (1997).
13. A. M. N. Ferguson, M. J. Irwin, R. A. Ibata, G. F. Lewis, and N. Tanvir, *AJ*, **124**, 1452–1463 (2002).
14. A. Ferrara, and E. Tolstoy, *MNRAS*, **313**, 291–309 (2000).
15. K. Freeman, and J. Bland-Hawthorn, *ARA&A*, **40**, 487–537 (2002).
16. J. S. Gallagher, G. J. Madsen, R. J. Reynolds, E. K. Grebel, & T. A. Smecker-Hane, *ApJ*, **588**, 326–330 (2003).
17. C. Gallart, A. Aparicio, G. Bertelli, and C. Chiosi, *AJ*, **112**, 1950–1968 (1996).
18. G. Gilmore, R. F. G. Wyse, and J. E. Norris, *ApJ*, **574**, L39–L42 (2002).
19. E. K. Grebel, *Reviews in Modern Astronomy*, **10**, 27–59 (1997)
20. E. K. Grebel, "Evolutionary Histories of Dwarf Galaxies in the Local Group", in *The Stellar Content of the Local Group*, edited by P. Whitelock & R. Cannon, IAU Symp. 192, Astronomical Society of the Pacific, San Francisco, 1999, pp. 17–38
21. E. K. Grebel, "The Star Formation History of the Local Group", in *Star formation from the small to the large scale*, edited by F. Favata, A.A. Kaas, & A. Wilson, 33rd ESLAB Symposium, ESA SP 445, ESA, Noordwijk, 2000, pp. 87–98
22. E. K. Grebel, *A&SSS*, **277**, 231–239 (2001).
23. E. K. Grebel, "The Evolutionary History of Local Group Irregular Galaxies", in *Origin and Evolution of the Elements*, edited by A. McWilliam & M. Rauch, Carnegie Observatories Astrophysics Series, Vol. 4, Cambridge University Press, Cambridge, 2004, pp. 237–257
24. E. K. Grebel, and P. Guhathakurta, *ApJ*, **511**, L101–L105 (1999).
25. E. K. Grebel, and P. B. Stetson, "Ground-based and WFPC2 Imaging of Fornax: Spatial Variations in Star Formation History", in *The Stellar Content of the Local Group*, edited by P. Whitelock & R. Cannon, IAU Symp. 192, Astronomical Society of the Pacific, San Francisco, 1999, pp. 165–169
26. E. K. Grebel, and J. S. Gallagher, *ApJ*, **610**, L89–L92 (2004).
27. E. K. Grebel, J. S. Gallagher, and D. Harbeck, *AJ*, **125**, 1926–1939 (2003).
28. C. J. Grillmair, et al., *AJ*, **115**, 144–151 (1998).
29. D. Harbeck, et al., *AJ*, **122**, 3092–3105 (2001).
30. D. Harbeck, J. S. Gallagher, and E. K. Grebel, *AJ*, **127**, 2711–2722 (2004).
31. D. Harbeck, J. S. Gallagher, E. K. Grebel, A. Koch, and D. B. Zucker, *ApJ*, in press (2005).
32. J. A. Holtzman, et al., *AJ*, **113**, 656–668 (1997).
33. D. Hunter, *PASP*, **109**, 937–950 (1997).
34. R. Ibata, G. Gilmore, and M. J. Irwin, *Nature*, **370**, 194 (1994).
35. R. Ibata, M. Irwin, G. Lewis, A. M. N. Ferguson, and N. Tanvir, *Nature*, **412**, 49–52 (2001).
36. C. Ikuta, and N. Arimoto, *A&A*, **391**, 55–65 (2002).
37. H. Jerjen, B. Binggeli, and K. C. Freeman, *AJ*, **119**, 593–608 (2000).
38. I. D. Karachentsev, and V. E. Karachentseva, *A&A*, **341**, 355–356 (1999).
39. I. D. Karachentsev, V. E. Karachentseva, A. A. Suchkov, and E. K. Grebel, *A&AS*, **145**, 415–423 (2000).

40. I. D. Karachentsev, et al., *A&A*, **389**, 812–824 (2002a).
41. A. Klypin, A. V. Kravtsov, O. Valenzuela, and F. Prada, *ApJ*, **522**, 82–92 (1999).
42. T. Lejeune, and J. Fernandes, *Observed HR Diagrams and Stellar Evolution*, edited by T. Lejeune & J. Fernandes, ASP Conf. Proc., Vol. 274, Astronomical Society of the Pacific, San Francisco, 2002
43. L. Mayer, F. Governato, M. Colpi, B. Moore, T. Quinn, J. Wadsley, J. Stadel, and G. Lake, *ApJ*, **547**, L123–L127 (2001).
44. N. F. Martin, N. F., R. A. Ibata, M. Bellazzini, M. J. Irwin, G. F. Lewis, and W. Dehnen, *MNRAS*, **348**, 12–23 (2004a).
45. N. F. Martin, N. F., R. A. Ibata, B. C. Conn, G. F. Lewis, M. Bellazzini, M. J. Irwin, and A. W. McConnachie, *MNRAS*, **355**, L33–L37, (2004b).
46. C. Mastropietro, B. Moore, L. Mayer, J. Wadsley, and J. Stadel, *MNRAS*, submitted (astro-ph/0412312), (2005).
47. M. Mateo, *ARA&A*, **36**, 435–506 (1998).
48. Y. Momany, S. R. Zaggia, P. Bonifacio, G. Piotto, F. De Angeli, L. R. Bedin, and G. Carraro, *A&A*, **421**, L29–L32 (2004).
49. B. Moore, S. Ghigna, F. Governato, G. Lake, T. Quinn, J. Stadel, and P. Tozzi, *ApJ*, **524**, L19–L22 (1999).
50. H. L. Morrison, P. Harding, D. Hurley-Keller, and G. Jacoby, *ApJ*, **596**, L183–L186 (2003).
51. H. J. Newberg, et al., *ApJ*, **569**, 245–274 (2002).
52. M. Odenkirchen, et al., *ApJ*, **548**, L165–L169 (2001).
53. M. Odenkirchen, et al., *AJ*, **126**, 2385–2407 (2003).
54. C. Palma, S. R. Majewski, M. H. Siegel, R. J. Patterson, J. C. Ostheimer, and R. Link, *AJ*, **125**, 1352–1372 (2003).
55. S. Pedraz, J. Gorgas, N. Cardiel, P. Sánchez-Blázquez, and R. Guzmán, *MNRAS*, **322**, L59–L63 (2002).
56. M. E. Putman, et al., *Nature*, **394**, 752–754 (1998).
57. M. Richer, M. L. McCall, & G. Stasinska, *A&A*, **340**, 67–76 (1998).
58. H. J. Rocha-Pinto, S. R. Majewski, M. F. Skrutskie, J. D. Crane, and R. J. Patterson, *ApJ*, **615**, 732–737 (2004).
59. G. A. Rutledge, J. E. Hesser, and P. B. Stetson, *PASP*, **109**, 907–919 (1997).
60. J. L. Sérsic, *Atlas de Galaxias Australes*, Obs. Astron. de Córdoba, Córdoba, 1968.
61. M. D. Shetrone, P. Côté, and W. L. W. Sargent, *ApJ*, **548**, 592–608 (2001).
62. T. A. Smecker-Hane, P. B. Stetson, J. E. Hesser, and M. D. Lehnert, *AJ*, **108**, 507–513 (1994)
63. H. Susa, and M. Umemura, *ApJ*, **610**, L5–L8 (2004).
64. K. Tassis, T. Abel, G. L. Bryan, and M. L. Norman, *ApJ*, **587**, 13–24 (2003).
65. E. Tolstoy, and A. Saha, *ApJ*, **462**, 672–683 (1996).
66. M. Tosi, L. Greggio, G. Marconi, and P. Focardi, *AJ*, **102**, 951–974 (1991).
67. S. van den Bergh, *ApJ*, **428**, 617–619 (1999).
68. S. van den Bergh, *A&ARv*, **9**, 273–318 (1999).
69. S. van den Bergh, *The Galaxies of the Local Group*, Cambridge Astrophysics Series Vol. 35, Cambridge University Press, Cambridge, 2000.
70. R. P. van der Marel, *AJ*, **122**, 1827–1843 (2001).
71. F. van Leeuwen, J. D. Hughes, and G. Piotto, *ω Centauri: A Unique Window into Astrophysics*, edited by F. van Leeuwen, J. D. Hughes, & G. Piotto, ASP Conf. Ser. 265, Astronomical Society of the Pacific, San Francisco, 2002.
72. K. A. Venn, M. Irwin, M. D. Shetrone, C. A. Tout, V. Hill, and E. Tolstoy, *AJ*, **128**, 1177-1195 (2004).
73. M. I. Wilkinson, J. T. Kleyna, N. W. Evans, G. F. Gilmore, M. J. Irwin, and E. K. Grebel, *ApJ*, **611**, L21–L24 (2004).
74. A. B. Whiting, G. K. T. Hau, and M. Irwin, *AJ*, **118**, 2767–2774 (1999).
75. A. B. Whiting, G. K. T. Hau, and M. Irwin, *ApJS*, **141**, 123–146 (2002).
76. R. F. G. Wyse, G. Gilmore, M. L. Houdashelt, S. Feltzing, L. Hebb, J. S. Gallagher, and T. A. Smecker-Hane, *NewA*, **7**, 395–433 (2002).
77. B. Yanny, et al., *ApJ*, **588**, 824–841 (2003).
78. D. Zaritsky, J. Harris, E. K. Grebel, and I. B. Thompson, *ApJ*, **534**, L53–L56 (2000).
79. Zucker, D. B., et al., *ApJ*, **612**, L121–L124 (2004a).
80. Zucker, D. B., et al., *ApJ*, **612**, L117–L120 (2004b).

Extragalactic Cataclysmic Variables

Michael M. Shara* and James D. Neill†

*American Museum of Natural History, 79th and Central Park West New York, NY, 10024
†Department of Physics and Astronomy, University of Victoria, Elliott Building, 3800 Finnerty Road, Victoria, BC, V8P 5C2, Canada

Abstract.
SALT is uniquely poised to make major inroads in the study of extragalactic cataclysmic variables (CVs) - novae and dwarf novae. The ability to search an external galaxy for erupting CVs night after night, for months at a time, AND to obtain confirmatory spectra within a night of discovery is unique and invaluable. We present several examples of multi-week to multi-month searches for extragalactic CVs with 1 to 4 meter-class telescopes. In particular, we have detected the first erupting dwarf novae in the LMC and placed a lower limit on the number of CVs in that galaxy. We have also observed the Local Group dwarf ellipticals M32 and NGC 205 in their entirety every clear night over a 4.5 month interval. In this survey we discovered one nova each in M32 and NGC 205, far more than previous nova surveys led us to expect. A similar search in M81 again reveals more novae than expected, and demonstrates, conclusively, that novae are predominantly a bulge population in spiral galaxies. Finally we report the detection of intergalactic tramp novae in the Fornax cluster, and emphasize that these are valuable tracers of stars stripped from their hosts during galaxy harassment. The insights gained during these preliminary studies illustrate how valuable SALT campaigns on extragalactic CVs will be.

INTRODUCTION

Cataclysmic Variables (CVs) beyond the disk and bulge of the Milky Way are potentially important as tracers of close binary stars in stellar environments very different from that near the Sun. As such, they are both standard candles (classical novae) and tests of our ability to predict close binary evolution (dwarf novae and classical novae) in galaxies very different from our own.

Dwarf novae (DN) were unknown outside the Milky Way until our recent detection in the LMC - see below - but may well number in the millions in all large spiral and elliptical galaxies. If we can use SALT and other 8 meter class telescopes to find significant extragalactic samples of DN we can do important comparisons concerning the eruption frequencies, amplitudes and binary orbital periods as functions of parent galaxy metallicity and age.

Current estimates of the bulk nova rates in galaxies such as M31 [11] of $\sim$37 yr^{-1} imply that in a few years the number of observed novae could be quite large for such a system. Novae represent an important complement to the increasing data on extragalactic X-ray binaries afforded by *Chandra* and *XMM*. The brightness of the nova outburst (M_V of -6 to -10 at maximum) betrays their presence to beyond the Virgo cluster with current telescopes. By observing extragalactic novae it is possible to trace the frequency and distribution of the close binaries that produce them in many extragalactic environments, thus allowing an exploration of close binary populations and the factors

CP752, *Stellar Astrophysics with the World's Largest Telescopes: First International Workshop*
edited by J. Mikołajewska and A. Olech

that influence their formation.

The normalized nova rate versus the luminosity of the host galaxy has been studied over the years [3, 10, 5], but systematic effects continue to dominate the published nova rates. We have found that nova rates are subject to biases that tend to underestimate the bulk rate for a given galaxy [7]. Most severe of these biases is the one imposed by telescope scheduling, which, until recently, provided only short, widely spaced runs for sampling nova rates in external galaxies.

We and others are attempting to overcome this bias by using dedicated telescopes to observe the target galaxy in its entirety, every clear night for many months. We will focus here on a few of our own surveys, with the goal of emphasizing how important queue-scheduled SALT-like telescopes can be in similar future studies.

OBSERVATIONS

Full details about each survey are given in our journal papers, but one factor is always in common - few telescope allocation committees have been far-sighted enough to allocate weeks or months to a single project (we are most grateful to those few that did!). In particular, we used the privately-owned Calypso telescope for over 6 months to search the galaxy M81 for classical novae. We used the Tenagra 0.8m telescope for our observations of the local group dwarfs M32 and NGC 205 over a total of 137 days. A two week-long dark observing run on the CTIO 4 meter enabled us to finally detect 6 likely LMC DN - something that SALT will do far better and far faster. Finally, a multi-year survey of the Fornax Cluster - again with the CTIO 4 meter - led to the first detections of intergalactic tramp novae, tracers of stars stripped from their host galaxies.

In all our surveys, images were visually and digitally blinked against each other to detect changing point sources. Nova and dwarf nova candidates were required to be observed in at least two epochs *and* to be missing on an epoch of sufficient depth coverage to confirm their transient nature. For the regions of our target galaxies near the nuclei, where the intensity gradient made detection more difficult, we used the spatial filtering technique described in Neill & Shara [7], allowing us to detect novae to within about 10" of the nuclei of M32, NGC 205 and M81.

LARGE MAGELLANIC CLOUD DWARF NOVAE

The eruptions of about 30 LMC classical novae have been well documented. DN and classical novae are, of course, intimately related: both are mass-transferring red dwarf-white dwarf binaries. All dwarf novae must eventually give rise to nova explosions, so DN MUST exist in the LMC. Until our survey, none were known.

To maximize our chances of detecting erupting LMC DN we requested the longest dark run with the largest telescope (the 4-m)and widest field imager (Mosaic2 8K $\times$ 8K) available in 1999 at CTIO. We were awarded the dark nights of 1999 December 2, 4, 6, 8, 10 and 12. All nights were clear, with seeing mostly in the range 1.0 to 1.8 arcsec. Because DN are usually remarkably blue (U-V ~ -0.7 is a good rule-of-thumb; though much redder values do occur Bailey [1]) we opted to image our chosen LMC field in U

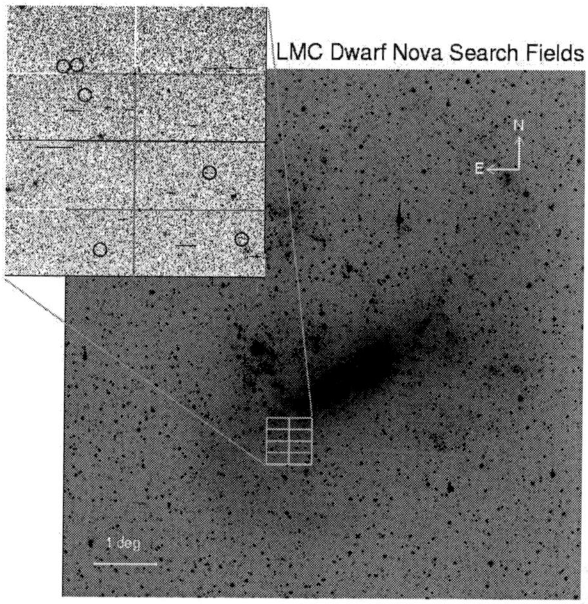

FIGURE 1. Field of the LMC imaged in our search for erupting dwarf novae, shown as an insert over the entire LMC. The scale bar on the bottom left is 1 degree long. N is up, E is left in the image.

and V filters.

The LMC field we chose is centered at RA (J2000) 05:33:36.7 and DEC (J2000) -70:33:44. An overall view of the location of this field within the LMC is shown in Figure 1, with a magnified view of the eight CCD fields superposed. The field is at the southeastern end of the LMC bar, and crowding is significant. The large surface density of stars hampered efforts to detect faint DN, but was critical to maximize the chances of detecting at least a few bright erupting DN.

CANDIDATE DWARF NOVAE

The dwarf-nova candidates were ultimately found by blinking rapidly through the best four of the six nightly frames to look for any subtle changes in the appearance of the stellar field. Dwarf novae should rise from invisibility to easily detectable in the 48 hours between observations.

To be recognized as a candidate, a variable had to be visible in both the U and V medianed frames of at least one of the six nights, and to have varied by over one magnitude in both filters over this timeframe. This selection method initially revealed 14 dwarf nova candidates from the entire dataset. Inspection of each night's individual images confirmed that 7 candidates were likely eclipsing binaries, and one candidate was so crowded that we cannot say for sure that it is a real variable. We were thus left with six good candidate erupting DN. We define a good candidate as one that is seen in

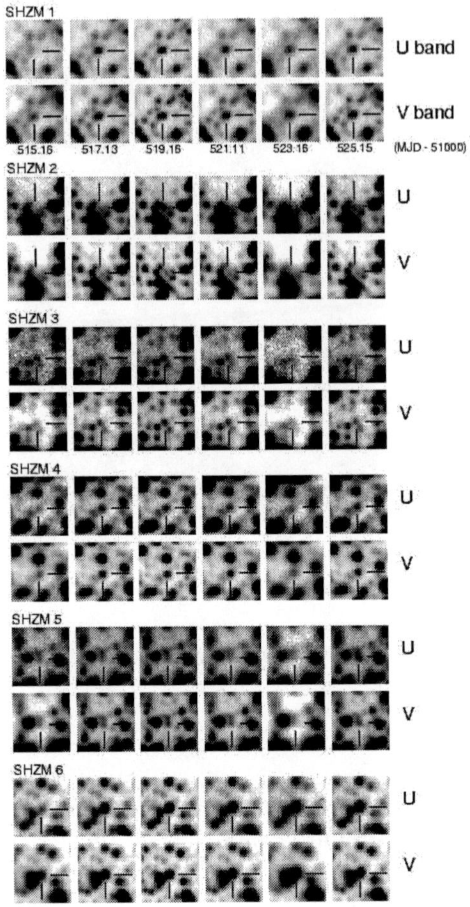

FIGURE 2. Nightly medians of images in U and V bands of the six candidate erupting dwarf novae in the LMC.

both U and V on at least one night; on all images in each filter of that night; and whose variability characteristics are consistent with those of at least one well-studied Galactic dwarf nova (see next section). The nightly images of the six DN candidates are shown in Figure 2, and their photometry is presented in Figure 3.

Simulations of eruptions of several well known and characterized Galactic dwarf novae, artificially placed in the actual LMC data, were carried out in a relatively uncrowded but otherwise representative patch of the LMC field. The images and photometry from these simulations are shown in Figures 4 and Figure 5, respectively. We regard the overall agreements between observations and simulations as strong, though not yet definitive. The brightness and variability behaviors of our candidates are consistent with those expected of luminous erupting dwarf novae in the LMC. We have considered, as possible imposters, chance superpositions of background supernovae or classical novae, gamma

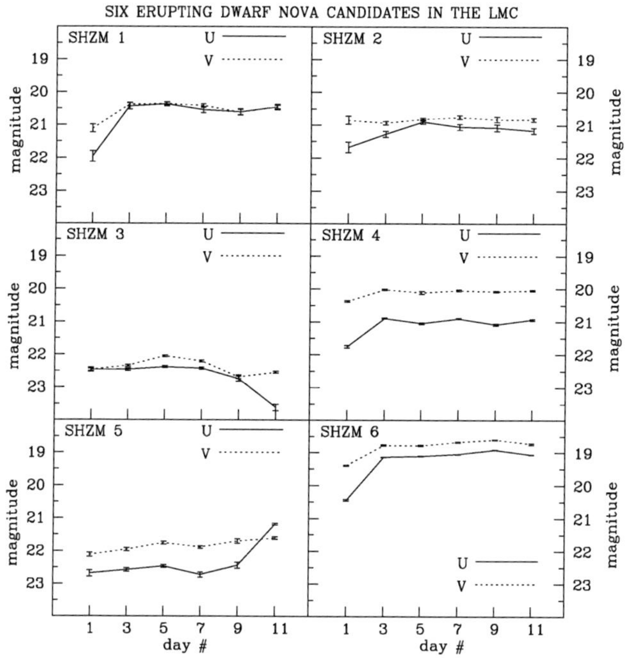

FIGURE 3. *U* and *V* light curves of the six dwarf nova candidates shown in Figure 2.

ray bursts (GRB), microlensing events or Milky Way variables along the line of sight to the LMC. All of these possibilities have been ruled out.

To be absolutely certain of these objects identities will require challenging follow-up observations. Four possible confirmation techniques are the following:

1. Spectra near quiescence (to demonstrate the presence of Balmer emission lines) would be definitive proof, but the DN are then expected to be near 25th magnitude... beyond the likely capability of even an 8 meter telescope in such crowded fields.
2. Imagery every night or two for several months with a 4-8 meter class telescope should reveal repeated eruptions separated by weeks to months. This can be done by SALT.
3. Ultraviolet imagery of the fields of the six candidate dwarf novae with the Hubble Space Telescope (HST) might reveal UV-bright objects, as the spectra of most CVs rise sharply into the UV. However, the significant internal reddening in the LMC [6] might mask this UV rise.
4. Several hours of time-resolved UV or optical photometry, again with HST, might reveal the flickering and/or orbital modulation characteristic of CVS.

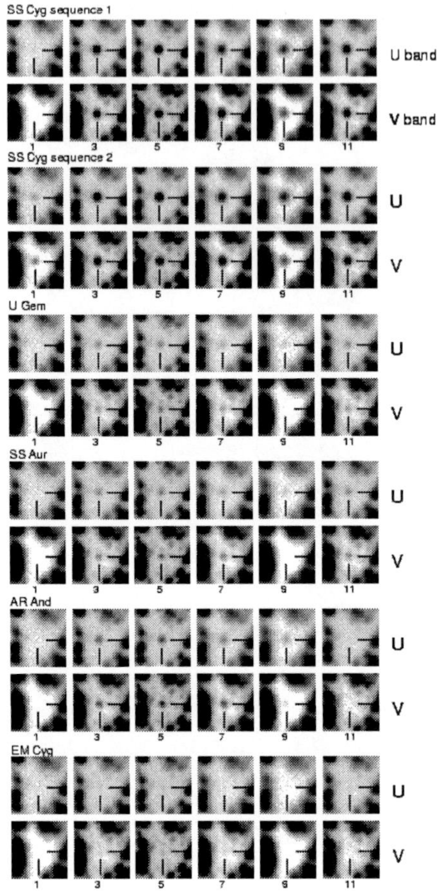

FIGURE 4. Simulated images of erupting Galactic dwarf novae placed in the LMC. See text for details.

Expected Dwarf Nova Populations in the LMC

Space density estimates of CVs near the Sun suggest 10^{-5} starspc^{-3} [9]. Space density estimates of all types of stars in the solar neighborhood yield 10^{-1} stars pc^{-3}, suggesting that about 1 CV exists in the Galaxy, and probably in the LMC, for every 10^4 stars. The LMC displays a V-band luminosity of -18.1, corresponding to a mass of $10^{10} M_\odot$ and a population of a few times 10^{10} stars). If the Galaxy and LMC manufacture CVs with similar efficiencies and rates then we estimate a total LMC CV population today of a few times 10^6 objects. About half of known CVs are DN which suggests a total LMC DN population of order 10^6. These DN are spread across the 10×10 degree surface of the LMC. Erupting classical novae are observed to be distributed quite uniformly across the face of the LMC, consistent with them belonging to an old

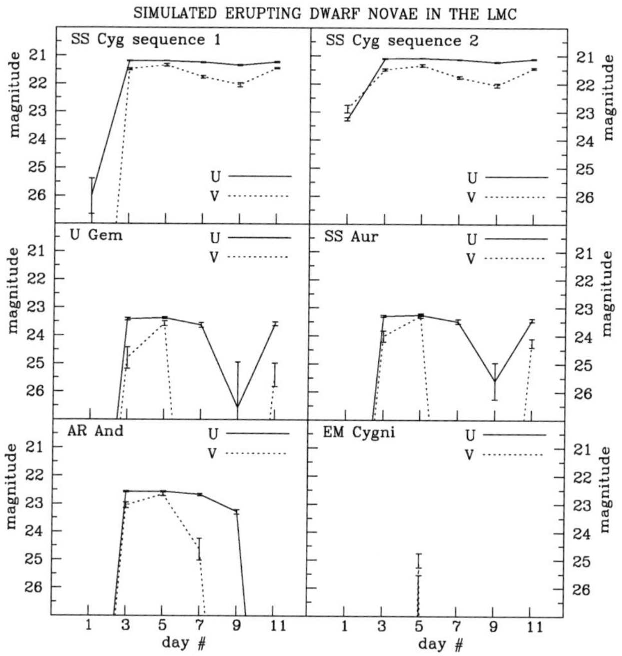

FIGURE 5. Simulated light curves of the dwarf novae of Figure 4.

population. A surface density of 10,000 DN per square degree across the LMC is thus expected.

Detections versus Expected Detections

The size of our field of view was 1362 arcmin^2 = 0.38 degree^2, which should include $0.38 \times 10000 = 3800$ DN. The average time between eruptions for 21 well studied Galactic DN is 29 days [12]. The length of our observing run (11 nights) suggests that any DN erupting on the nine nights between nights 2 and 10, inclusive, could have been detected (if it became sufficiently luminous). Hence $(9/29) \times 3800 \sim 1200$ DN eruptions occurred in our field of view that were (at least in principle) detectable.

The fact that we detected at most six DN when (an admittedly simple) model predicts 120 suggests that many eruptions were too faint to be recorded. This is in accord with the model light curves of LMC DN. If our six candidates are all bona fide erupting DN in the LMC this suggests that 95% of all DN are less luminous than SS Cyg.

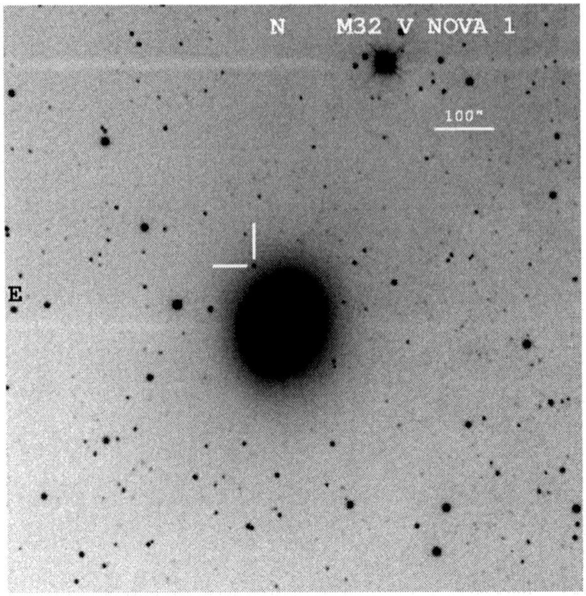

FIGURE 6. Tenagra V image of M32 Nova 1 on JD 2453009.62. North is up and East to the left. The nova is indicated by the two lines at right angles. A scale bar of 100 arcseconds is also shown.

AN M32 NOVA

A 137 day-long search of NGC 205 and M32 yielded one erupting classical nova in each of these two dwarf elliptical companions to the Andromeda galaxy. We concentrate here on the M32 nova, shown in outburst in Figure 6, while Figure 7 presents the calibrated light curve for this nova. The frame limits are plotted as short horizontal lines with downward pointing arrows, while the open points with error bars are the nova observations. A simple linear fit was made to the decline portion of the light curve in V to calculate the decline rate in m_V day^{-1}. A unique aspect of this nova is that we were able to obtain its spectrum AND to search the HST archives to constrain the brightness of its progenitor.

Verifying the M32 Nova

Our spectral observations of M32 Nova 1 confirm that it is indeed a nova. They show broad Hα and Hβ in emission with a velocity of expansion of 640 km s^{-1}, typical of a slow classical novae. Figure 8 shows the spectrum taken on JD 2453017.60, roughly ten days after maximum. Emission lines of Fe II can be seen just red-ward of Hβ which is also consistent with it being a typical slow nova. The spectra taken five days later is nearly identical, an indication of the slow spectral evolution of this nova.

Because M32 is superposed on the outer disk of the larger M31, we also wanted to

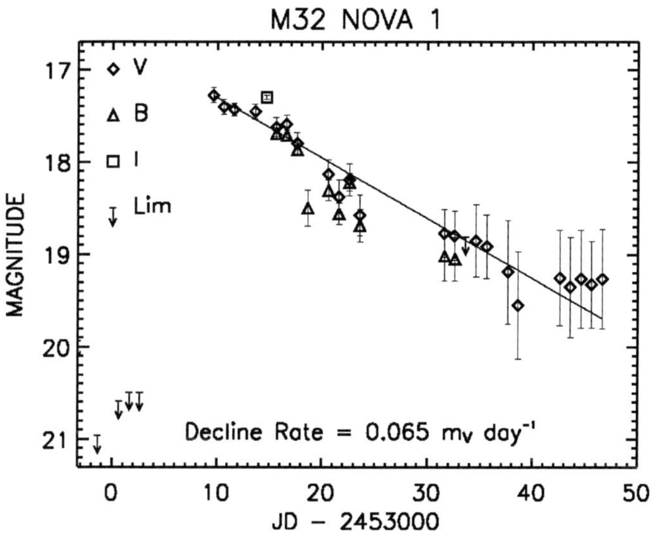

FIGURE 7. Light curve of M32 Nova 1 in V, B, and I-bands. V-band points are indicated by diamonds, B-band by triangles, and the I-band point by the square. Frame limits are indicated by the short line with a downward pointing arrow. The decline rate of 0.065 m_V day^{-1} was determined from an error-weighted linear fit (shown by the thin line) to the V-band points.

confirm that this nova did not originate in M31. We measured the systemic heliocentric radial velocity of the nova from the Hα lines and found it to have a velocity of -170 ± 6 km s^{-1}. This is consistent with the radial velocity of M32 (-200 ± 6 km s^{-1}) if we account for the velocity dispersion of M32 at a radius of 78". The radial velocity of M31 is -297 ± 1 km s^{-1}. The part of M31's disk upon which M32 is superposed is where most of the disk's rotation velocity would be transverse to the line of sight and therefore could not make up the difference in velocity between M31 and the nova. This, combined with the proximity of the nova to the nucleus of M32, argues against it originating in M31.

In addition to these spectral confirmations, we also put a limit on the nova outburst amplitude. To do this, we used the *HST* WFPC2 observations in V (F555W) taken nearly a decade prior to outburst. We also used the I (F814W) and F1042M images to verify that the progenitor was not a bright red variable. The nova position in all these filters was in the WFPC2 chip WF3.

Figure 9 shows the region in the WF3 V image surrounding the nova, with a field of view of 3 arcseconds on a side. We do not convincingly detect the progenitor of M32 Nova 1. The star indicated by the letter A is at V of 25.5 and I of 24.5, but is well outside the position error circle defined by the RMS of the nova positions. We estimate that the progenitor of this nova was fainter than V of 26.0 at the epoch that these data were

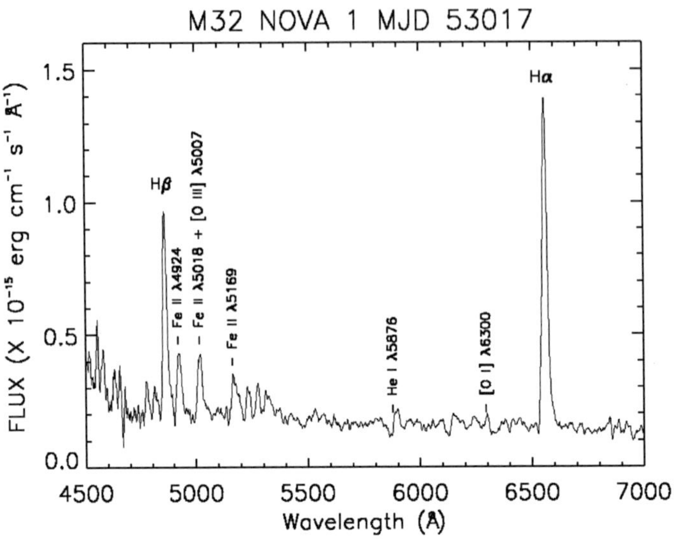

FIGURE 8. Spectrum of M32 Nova 1 taken on JD 2453017.60, 8 days after discovery. The broad H emission lines confirm it as a classical nova. The Hα line has a half width at half intensity of 640 km s^{-1}, at the low end of the velocity range (typically 300 to 3000 km s^{-1}) for classical novae. The strength of the Fe II lines and the weakness or absence of the He and N lines is consistent with the slow nova classification. A spectrum taken on JD 2453022.60, 5 days later, is nearly identical to the spectrum presented here, illustrating the slow spectral evolution of this nova. Note that the He I line (λ5876) has a P Cygni profile and the Hα line also shows a hint of absorption on the blue side.

taken. This gives an amplitude of at least 8.7 magnitudes in V for the nova outburst. The WFPC2 I and F1042M images were well registered with the V image and nothing was detected in them at the position of the nova. We conclude, therefore, that M32 Nova 1 could not be a bright red variable and is, indeed, a nova.

THE M32 NOVA RATE

A raw bulk nova rate for a galaxy can be obtained simply by dividing the observed number of novae that erupted during the survey by the time covered. This gives $R = 1/0.355\text{yr} = 2.82\ \text{yr}^{-1}$ for M32. A more sophisticated Monte Carlo approach yields a bulk nova rate for M32 of $2^{+2.4}_{-1.0}\ \text{yr}^{-1}$. This bulk nova rate is much higher than predicted by extrapolating a constant Luminosity Specific Nova Rate (LSNR) determined at high luminosities to low luminosities.

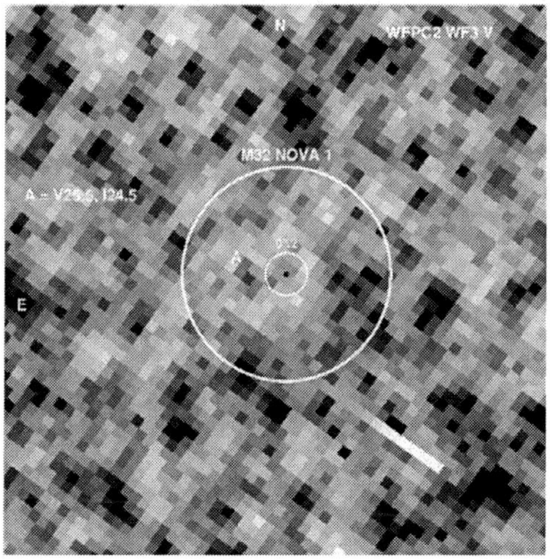

FIGURE 9. *HST* WFPC2 WF3 V-band image of the region around M32 Nova 1 taken on JD 2449622.62. North is up and East to the left. The circles are centered on the position of M32 Nova 1. The larger circle has a radius of 1" and the smaller a radius of 0.2. The positional error of M32 Nova 1 is 0.1 WF pixels or 0.01 and is represented as the black dot at the center of the figure. Star A has a V magnitude of 25.5, but is well outside the 1σ error circle of M32 Nova 1. The progenitor of M32 Nova 1 is fainter than V of 26.0, implying an outburst amplitude of over 8.7 V magnitudes.

NOVAE IN M81

Initial results of our extensive searches for novae in M81 have already appeared in the literature, [7] so we only present a brief update here. Recent, as-yet-unpublished new M81 novae have been added to those already published (see Figure 10), and strengthen both our rate and distribution conclusions. In brief, our extensive and densely time-sampled surveys of the novae in M81 indicate that they are about 90% bulge population...but spiral arm novae certainly exist. This is abundantly clear from from Figure 10. Our rate for novae in M81 is at least 30 per year.

TRAMP NOVAE IN THE FORNAX CLUSTER OF GALAXIES

A fundamental problem in understanding the evolution of galaxy clusters and their constituent galaxies is the importance of dynamical processes such as tidal stripping. Theoretical models of cluster evolution have considered collisions, mergers, dynamical friction and tidal stripping in order to explain the galaxy morphology-density relation and the presence of luminous halos around supergiant D galaxies. It is generally accepted that tidal stripping will remove the more loosely bound stars in the outer envelopes of galaxies. However, the efficiency of tidal stripping is not well understood, because of uncertainties in the velocities, the mean tidal field, the distribution of dark matter, and the orbits of stars in the outer envelopes of the stripped galaxies. Typical theoretical es-

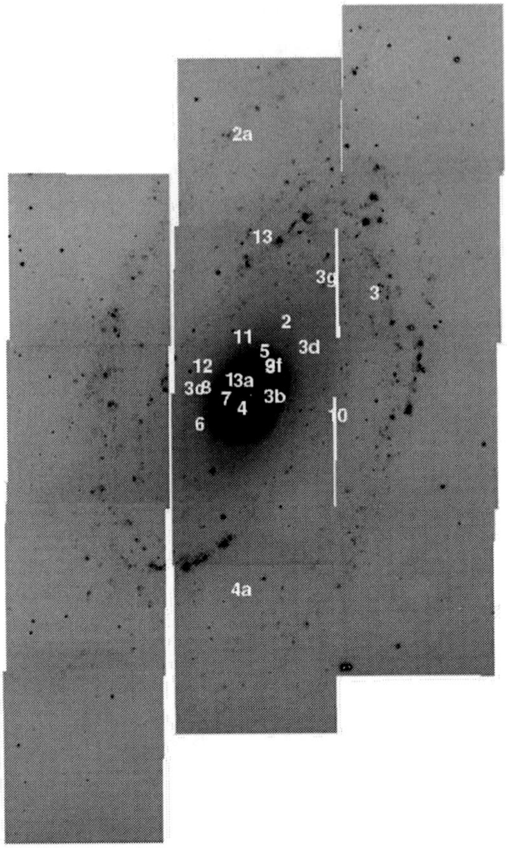

FIGURE 10. The distribution of novae discovered in M81 during a 6 month-long, nightly survey.

timates indicate that about 20% of the total cluster light should be in the intergalactic medium and that the amount of ICL should increase with time. Even with CCDs, detecting the ICL directly is notoriously difficult and fraught with uncertainties such as the contributions from overlapping galaxy halos. An independent means of measuring the ICL, not subject to the same problems as measuring the extremely low surface brightness diffuse emission is needed. Methods using planetary nebulae (PNe), intergalactic red-giant-branch stars have been used in Virgo and Fornax. Novae typically range in brightness from $M_V = -6$ to -10, while the bright end of the PN luminosity function reaches $M_{(5007)} = -4.5$, while red-giant-branch stars typically reach $M_V = -3$. Novae can therefore be detected more easily in a given cluster, or be observed to greater distances than PNe or red giant stars. The transient nature of novae and their Hα brightness help eliminate the contamination suffered by PNe or red-giant-branch stars due to background emission line objects or unresolved compact galaxies. Another appealing property of novae is that the rate of novae per unit mass is high enough such that with a deep enough survey, one can expect to accrue a statistically interesting number of novae in each epoch, thus increasing the accuracy of the ICL measurement with each new epoch.

We acquired six epochs of survey data of the central 30 arcminutes^2 of the Fornax cluster. Fornax is nearby and has a fairly high surface density of galaxies in the core, increasing the likelihood of tidal stripping. At the distance of Fornax, (m - M) = 31.35, novae will have an average m(peak) = 23.35 (the brightest novae will have peak brightnesses of m = 21.35), and are easily detectable.

Novae are very bright in H-alpha (B - Hα = 0.5 when the nova erupts, and B - Hα = 3 shortly thereafter). Furthermore, novae remain bright in H-alpha for an extended period of time (months), providing a useful means of discovering them. To a limiting magnitude of M(Hα) = -7.5 the mean Hα visibility lifetime is 3 months (Ciardullo et al. 1990). We estimated that at any given epoch there should be about 50 novae in eruption (with visible H-alpha emission) in galaxies in the central square degree of the Fornax cluster.

Observations

The observations that went into this study are a diverse set acquired over the span of more than a decade. The central 30' of Fornax, centered on RA 03:38:13.72, Dec -35:32:52.2 (J2000), was surveyed. This area includes the central dominant (cD) galaxy NGC 1399 and three other elliptical galaxies: NGC 1404, NGC 1387, and NGC 1389. This region was imaged twice in 1993, once in 1995, for five nights in 1999, and once each in 2003 and 2004. The data were acquired during programs for other projects and during engineering time or director's discretionary time.

Nova Detection

Classical novae erupt with amplitudes of up to 20 magnitudes, reaching absolute magnitudes as high as $M_B = -10$ and fading back to their pre-outburst brightness in a matter of weeks or months (Warner 1995). Novae in the Fornax cluster are therefore invisible before outburst and fade back to invisibility after outburst on time scales of less than a year. Thus, our primary criterion for detecting novae is that they be transient on this timescale. For the data used in this study this means that they must only appear in only one year's images. If a variable object appears in images from more than one year, it is most likely some other kind of variable source: a foreground periodic, or semi-periodic variable, or a background, unresolved active galaxy nucleus (AGN).

Classical novae also exhibit strong Hα emission in outburst, providing the next criterion: the candidate should be Hα bright. Data from 1993 and 1999 were taken both through H-alpha and red continuum filters allowing us to produce Hα minus red continuum subtracted images. These images were used to test candidates from 1993 and 1999 against this criterion. The primary method for implementing these criteria consisted of performing a spatial registration of the different epochs, and then blinking them against each other. We required that any candidate nova be visible on each of the

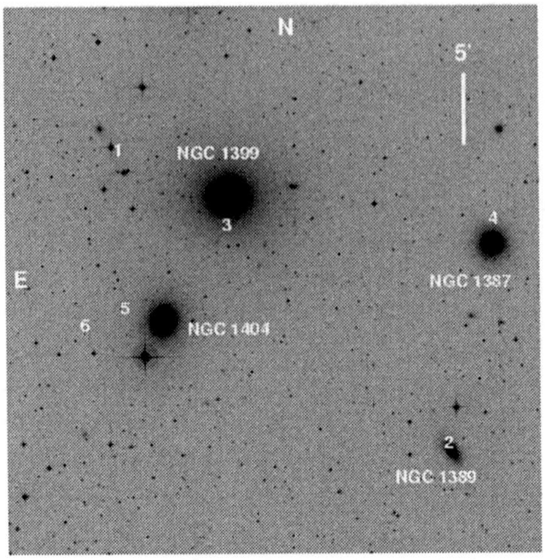

FIGURE 11. Locations of the 6 novae discovered in the Fornax cluster overplotted on a DSS image. Novae are indicated by their identification number. North and East are at their usual orientation as labeled. The field of view is 38' on a side. The galaxy identifications are indicated, along with a 5' scale bar in the upper right corner. Note the positions of Nova 1 and Nova 6, strong intracluster tramp candidates.

individual exposures for the epoch in which it was detected. This eliminated cosmic ray hits from our candidate list. This method worked well for candidates outside of the galaxies. For candidates in the galaxies, we employed the spatial median filter technique described in Neill & Shara (2004) with a box size of 17 pixels. Figure 11 shows the locations of the detected nova candidates on an image from the Digitized Sky Survey 2 covering the central 38' of the Fornax cluster, corresponding to our survey area. The nova candidates are labeled by number and the galaxy identifications are also indicated.

Are These Variables Really Novae?

Could these six candidates be some other kind of variable and not novae? Without spectra, it is impossible to be certain. All of our candidates are transient according to the criterion discussed above, thus making periodic or semi-regular variables unlikely contaminators. Four of the six candidates (Novae 2, 3, 4, and 6) are Hα bright and transient, virtually ruling out other possible variables. Other types of variables that are transient include background AGN, X-ray bursts (GRBs), supernovae (SNe), and microlensing events. We had to resort to statistical arguments to eliminate each of these classifications for Novae 1 and 5.

TRAMP NOVA PROPERTIES

We were fortunate that three of the novae erupted during the five epoch run in 1999. The two measured decline rates (see Figure 8) of 0.183 m_B per day for Nova 3 and 0.205 m_B

per day for Nova 6 place both in the fast nova category. If we assume a distance modulus to the Fornax cluster of $(M - m)_0 = 31.35$, then our brightest B magnitude for Nova 3 is at M = -9.2. Typically, novae reach peak B magnitudes of M = -8.0, so Nova 3 is clearly a brighter-than-average nova (see, e.g., Figure 5.3 in Warner 1995). The flat shape of the Hα light curve indicates that we have caught this nova near Hα maximum and during the earliest phases of the B decline. The brightness of Nova 3 is consistent with its fast decline rate since decline rate is known to correlate with nova peak brightness. Given the decline rate of Nova 6, it is possible that this nova was at least as bright as Nova 3 at maximum, but that we are observing it in its later decline phase. The large uncertainty in the Hα magnitude for this nova does not allow us to compare its B and Hα magnitudes. If we assume the distance modulus to Fornax quoted above, then we have sampled a range in nova brightness from M = -6.6 (Nova 4) to M = -9.2 (Nova 3) and from $M_{H\alpha}$ = -7.2 (Nova 6) to $M_{H\alpha}$ = -10.2 (Nova 2). These basic properties show that we have sampled a significant range of B and Hα brightnesses.

THE SPATIAL DISTRIBUTION OF THE NOVAE

We deliberately limited ourselves to two simple questions about the spatial distribution of novae in the Fornax cluster. First, assuming that novae follow the bulge light in galaxies (Neill & Shara 2004), how many novae did we miss in the centers of the galaxies? Second, based on the light profiles of the galaxies, how many novae are associated with the cluster potential and not with any galaxy? Assuming that novae trace the intracluster light in a galaxy cluster, we calculate that the ICL contributes 3/(4.3 + 3) = 41% of the total light in the Fornax cluster. Our key result is that we have demonstrated for the first time that intergalactic tramp novae very likely exist in the Fornax cluster, and that they can be useful tracers of the ICL.

REFERENCES

1. Bailey, J. 1980 MNRAS, 190, 119
2. Ciardullo, R., Shafter, A. W., Ford, H. C., Neill, J. D., Shara, M. M., & Tomaney, A. B. 1990, ApJ, 356, 472
3. della Valle, M., Rosino, L., Bianchini, A., & Livio, M. 1994, AAP, 287, 403
4. della Valle, M. & Livio, M. 1998, ApJ, 506, 818
5. Ferrarese, L., Côté, P., & Jordán, A. 2003, ApJ, 599, 1302
6. Gochermann, J., & Schmidt-Kaler, T. 2002, AA, 391, 187
7. Neill, J. D. & Shara, M. M. 2004, AJ, 127, 816
8. Nowotny, W., Kerschbaum, F., Olofsson, H., & Schwarz, H. E. 2003, AAP, 403, 93
9. Patterson, J. 1998, PASP, 110, 1132
10. Shafter, A. W., Ciardullo, R., & Pritchet, C. J. 2000, ApJ,530, 193
11. Shafter, A. W. & Irby, B. K. 2001, ApJ, 563, 749
12. Szkody, P. & Mattei, J. 1984, PASP, 96, 988
13. Cataclysmic Variable Stars, Cambridge Astrophysics Series (Cambridge, New York: Cambridge University Press) 1995

Distance Indicators: A Connection Between Stellar and Extragalactic Astrophysics

Gary Wegner

Department of Physics & Astronomy, 6127 Wilder Laboratory, Dartmouth College, Hanover, NH 03755-3528, USA

Abstract.
Galactic distance indicators, particularly the Fundamental Plane and the Tully-Fisher relations are discussed. These have both been widely used to measure large-scale motions of galaxies in the local universe and to extract cosmological information. However these indicators can be calibrated using nearby galaxies such as those in the local group and modelled using detailed N-body simulations and observations of stellar populations. These considerations provide a bridge between distant and nearby extragalactic objects and their stars.

INTRODUCTION

Besides defining the cosmological distance ladder, the extragalactic distance indicators can provide much astrophysical information. One application is measuring cosmological peculiar motions, but this depends on the details of the structure, dynamics, and stellar populations of galaxies which produce these relations. These topics relate in general to the evolution of large-scale structure and galaxies and consequently stellar observations of nearby objects in the Milky Way and the Local Group provide complimentary information.

A nearby galaxy's peculiar velocity, V_{pec}, is the radial component of its motion above or below the redshift given by the smooth expansion of the universe and relates to the observed redshift, cz, by $cz = H_0 D + V_{pec}$. To find V_{pec}, requires a distance indicator yielding D and numerous investigators utilize diverse distance indicators to measure V_{pec}, *e.g.* 7S (Lynden-Bell et al. 1988), MKIII (Willick et al 1997), SFI (Giovanelli et al. 1997), EFAR (Wegner et al. 1999), ENEAR (Wegner et al. 2003), and SMAC (Hudson et al. 2004). Dekel (2000) and Zaroubi (2002) provide more comprehensive lists. The future use of the SZ effect and X-ray observations promise interesting results.

MOTIVATIONS

Whilst V_{pec} have become overshadowed by the successful CMB experiments, supernovae, the Sloan and 2df redshift surveys in obtaining cosmological information, important measurements remain possible, for using V_{pec} makes assumptions independant of the other studies. V_{pec} results from all forms of gravitating matter and hence probes the total large-scale structure. Density and velocity field reconstruction and determination of the power spectrum of density fluctuations (e.g. Strauss & Willick 1995; Dekel 1999;

CP752, *Stellar Astrophysics with the World's Largest Telescopes: First International Workshop*
edited by J. Mikołajewska and A. Olech

Zaroubi 2002) and earlier results established a low Ω universe and large mass concentrations such as the Great Attractor (Lynden-Bell et al. 1988). An independant method of determining the cosmic density parameter, Ω, using pairwise V_{pec} data and the BBGKY hierarchy that is empirically based and free of many assumptions has been described by Juskiewicz et al. (2000) and Feldman et al. (2003) which yields $\Omega = 0.30^{+0.17}_{-0.07}$.

The Fundamental Plane (FP) and Tully-Fisher (TF) relations, although not the most precise distance indicators, can be measured in large numbers and provide the most galaxies at relatively large distances for V_{pec} studies. The FP and TF intrinsic scatters exceed their measurement errors and so far no theoretical simulation reproduces their slopes and zero points. Although not surprising given the present state of knowledge, the question of the origin of the relations and their scatter and how to improve their accuracy remains. Finding an answer leads deeply into considerations of the origin of galaxies and the evolution of their stellar populations. Space does not permit discussing the observational techniques and corrections employed in their measurement, but the following briefly summarizes the two relations.

The Fundamental Plane (FP) Relation

The FP relation for nearby early-type galaxies is derived from fitting low z clusters and is typically of the form:

$$\log R_e = (1.22 \pm 0.09) \log \sigma + (0.336 \pm 0.013) < SB_e > +const.,$$

where R_e is the effective radius and σ and $< SB_e >$ denote the galaxy's central velocity dispersion and mean surface brightness within R_e (Saglia et al. 2001; Colless et al. 2001; Bernardi et al. 2003). The relation's resulting scatter indicates a typical distance error of about $\pm 15\%$ per object.

The FP does not depend strongly on environment (Lucey et al. 1991) despite the strong morphology-density relation. Chemical evolution, also discussed in the accompanying references places constraints on the evolution. Line strength measurements in clusters (Colless et al. 1999; Caldwell et al. 2003; Smith et al. 2004; Nelan et al. 2004) and in voids (Wegner & Grogin 2004) suggest similar results, although age differences might be expected. High-z clusters show systematic zero point shifts and larger scatter (Cf. Fritz & Ziegler 2004; Wuyts et al. 2004; van der Wel et al. 2004; van de Ven et al. 2003) attributable to age variations. The light profiles of the early-types also provide information through their light profiles. Fitted by an exponential disk and a Sersic $r^{\frac{1}{n}}$ bulge (de Jong et al. 2004),this is consistent with large bright E galaxies forming from merging of a few large spirals, although such mergers cannot create low L E's. Also dynamical studies indicate extended dark matter haloes (Thomas et al. 2004).

The Tully-Fisher (TF) Relation

The TF relation for late-type spirals relates absolute magnitude to the inclination corrected rotation width, W_0, and is typically of the form

$$M_I = (-7.68 \pm 0.13)(\log W_0 - 2.5) - (21.00 \pm 0.02).$$

Here an I-band relation is given (Giovanelli et al. 1997). The scatter is about ± 0.35 mag. per object and produces distance errors comparable to the FP. For galaxy clusters containing N galaxies, this is of course reduced by $\sqrt{N}$ for both the FP and TF relations.

The TF relation also changes with z. Ziegler et al. (2002, 2003), Böhm et al. (2004), Fuchs et al. (2004), Fritz & Ziegler (2004) report a shallower slope plus zero point and scatter changes, which they interpret as indicatative of a mass dependent M/L ratio evolution.

THEORETICAL INTERPRETATIONS

Many recent investigators attempt to explain the FP and TF. For the TF, Recent work (Cf. Abadi et al. 2003; Steinmetz & Navarro 1999, 2002; Balogh et al. 2000; Mo et al. 1998) typically use ΛCDM simulations to form spirals from mergers and accretion events. Although the TF slope and scatter are reproduced by hierarchical clustering in such a universe, the models cannot account for the zero point and the flatter slope reported at $z \sim 1$.

Several groups consider the FP as important for galaxies as the HR diagram is for stars (e.g. Bender et al. 1993). Numerical experiments indicate both dynamical and M/L evolution are factors in forming the FP (Cf. Dantas etal. 2003; Gonzales-Garcia & van Albada 2003; Borriello et al. 2003; Estigneeva et al. 2004). N-body simulations indicate that bright E's cannot form from mergers of dwarf galaxies resembling present day dwarfs. While reasonable models produce FP-like relations, they do not reproduce the observed FP. In summary, no generally accepted formation model yet exists for both relations.

BRIDGING NEAR AND FAR OBJECTS

The Milky Way and Local Group result from the same processes occuring at high z that have formed the TF and FP. Thus the same questions regarding their origin and the presence of dark matter present themselves. In addition, any explanation must account for the galaxies' chemical evolution. The Eggen et al. (1962) monolithic collapse idea needs serious modification in favor of the hierarchical picture, but the precise processes remain elusive. The recent delineation of debris trails indicate that the tidal disruption and devouring of satellite galaxies been in forming the galaxy's disk (Cf. Feast 2004; Ibata et al. 2001, 2003; Majewski et al. 2003; Majewski 2004). Grebel (2004a,b) discusses the evolutionary history of the Local Group. The Irr and dIrr galaxies which give a closeup record of objects forming with the larger galaxies are important concerning their dark matter content and enrichment histories, but present data are inconclusive about their metal enrichment and evolution histories. Thus questions about the formation of objects ranging from stellar to extragalactic objects overlap and one can anticipate that future observations with the SALT will contribute towards their solutions.

REFERENCES

1. Abadi, M. L. et al. 2003, ApJ, 591, 499
2. Balogh, M. L. et al., 2000, ApJ, 540, 113
3. Bender et al. 1993, ApJ, 411, 153
4. Bernardi, M. et al. 2003, AJ, 125, 1866
5. Böhm, B. L. et al. 2004, A&A, 420, 97
6. Borriello, A. et al. 2003, MNRAS, 341, 1109
7. Caldwell, N. et al. 2003, AJ, 125, 2891
8. Colless, M. et al. 2001, MNRAS, 321, 277
9. Dantas, C. C. et al. 2003, MNRAS, 340, 398
10. de Jong, R. S. et al. 2004, MNRAS, in press, astro-ph/0410121
11. Dekel, A. 2000, ASP Conf. Ser. 201, Cosmic Flows Workshop
12. Eggen, O. J. et al. 1962, ApJ, 136, 748
13. Estigneeva, E. A. et al. 2004, MNRAS, 349, 1052
14. Feast, M. 2004, in Sci. SALT Workshop Proc., vol 2
15. Feldman, H. et al. 2004, ApJ, 596, L131
16. Fritz, A. & Ziegler, B. L. 2004, astro-ph/0405078
17. Giovanelli, R. et al. 1997, AJ, 113, 53
18. Gonzales-Garcia, A. C. & van Albada, T. S. 2003, MNRAS, 342, L36
19. Grebel, E. K. 2004a, in Carnegie Obs. Ap. Ser. vol. 4, p. 237
20. Grebel, E. K. 2004b, these procedings
21. Hudson, M. J. et al. 2004, MNRAS, 352, 61
22. Ibata, R. et al. 2001, ApJ, 551, 294
23. Ibata, R. et al. 2003, MNRAS, 340, L21
24. Juskiewicz, R. et al. 1999, ApJ, 518, L25
25. Lucey, J. R. et al. 1991, MNRAS, 253, 584
26. Lynden-Bell, D. et al. 1988, ApJ, 326, 19
27. Majewski, S. R. et al. 2003, ApJ, 599, 1082
28. Majewski, S. R. 2004, these procedings
29. Mo, H. J. et al. 1998, MNRAS, 295, 319
30. Nelan, J. et al. 2004, Preprint
31. Saglia, R. P. et al. 2001, MNRAS, 324, 389
32. Smith, R. et al. 2004, AJ, 128, 1558
33. Steinmetz, M. & Navarro, J. F. 1999, ApJ, 513, 555
34. Steinmetz, M. & Navarro, J. F. 2002, New Astr. 7, 155
35. Strauss, M. A. & Willick, J. A. 1995, Phys. Rep. 261, 271
36. Thomas, J. et al. 2004, MNRAS, submitted
37. van de Ven, et al. 2003, MNRAS, 344, 924
38. van der Wels, A. 2004, ApJ, 601, L5
39. Wegner, G. et al. 1999, MNRAS, 305, 259
40. Wegner, G. et al. 2003, AJ, 126, 2268
41. Wegner, G. & Grogin, N. 2004, preprint
42. Wuyts, S. et al. ApJ, 605, 677
43. Zaroubi, S. 2002, astro-ph/0206052
44. Ziegler, B. L. et al. 2002, ApJ, 564, L69
45. Ziegler, B. L. et al. 2003, ApJL, 598, 87

Dynamics of Disrupting Milky Way Satellite Galaxies and Tidal Streams

S. R. Majewski*, R. R. Muñoz*, K. B. Westfall*,† and R. J. Patterson*

*Dept. of Astronomy, University of Virginia, Charlottesville, VA 22903-0818, USA
†Dept. of Astronomy, University of Wisconsin, Madison, WI 53706, USA

Abstract.
The nature of the Galactic dSph satellites remains controversial. Do they have extreme M/L? Are they tidally disrupting? Do they contribute significantly to the creation of the halo and halo substructure? Does mass follow light in these systems? Great insight into these questions will come from high quality radial velocity surveys over large areas, especially for dSph stars well outside the nominal King limiting radii. For most satellites, this work is the realm of 8-m class echelle spectroscopy, however, the density of suitable stars outside the tidal radius is low enough that little advantage is conferred by most multi-object spectrographic systems. Worse, the density of good candidates is so low compared to that of foreground Milky Way dwarf stars that clever pre-filtering is needed to make efficient use of telescopes like SALT for this type of science.

DWARF SPHEROIDAL ISSUES

More than a half-century since their discovery, the nearby dwarf spheroidal (dSph) galaxies still present us with a number of perplexing problems and apparent dilemmas concerning their nature and evolution. The era of 8 to 10-m class telescopes gives great promise that we will soon resolve many of these issues, since apertures of this size mean that high resolution spectroscopy of the constituent stars in these stellar systems are now brought within easy reach. We focus here primarily on dynamical aspects of the dwarf spheroidals and their connection to the Milky Way halo, but a host of additional problems relating to the chemical evolution and star formation histories of these systems are now being addressed with high S/N echelle spectroscopy of large statistical samples of stars within these galaxies (see contribution by Grebel to this proceedings).

Among the dynamical intrigues that dSph still pose are:

(1) *Why do* dSph *galaxies have the highest apparent dark matter fraction?* This is an especially serious discrepancy when compared with stellar systems of nearly similar scale — e.g., globular clusters, and dwarf ellipticals, which have $M/L <\sim 10$ (all M/L values here are in solar units). In contrast, Galactic dSphs apparently have M/L as high as ~ 100, while values even several times larger than this have recently been proposed for some of these systems (e.g., Draco: Kleyna et al. 2002, Łokas et al. 2004).

(2) *Does mass follow light in these systems?* New studies of velocities of dSph stars ever farther from the dSph centers are showing dispersions that continue to be large, with flat or even rising σ_v with radius (Kleyna et al. 2002, 2004; Majewski et al. 2004a; Westfall et al. 2004; Muñoz et al. 2004). One way to accommodate this trend is to embed the luminous dSphs within extended dark matter halos, with the implication of even

CP752, *Stellar Astrophysics with the World's Largest Telescopes: First International Workshop*
edited by J. Mikołajewska and A. Olech

larger total M/L in these systems (e.g., 440±240 for Draco according to Kleyna et al. 2002). However, more recently Muñoz et al. (2004) have had success accounting for these flat/rising velocity dispersion trends by attributing it to tidal disruption; in these circumstances, simple mass-follows-light models are found to give reasonably good descriptions of the dSph system.

(3) *What is the extent of tidal effects on dSph galaxies?* N-body models of dSph-like systems in Milky Way-like potentials suggest that even systems with reasonably large masses can suffer tidal disruption, and the unbound debris can be quite near to the center of the dSph to create the flat dispersion profile (e.g., Muñoz et al. 2004). Indeed, a flat σ_v with radius is exactly what is observed for the Sagittarius (Sgr) system from the center to hundreds of degrees down the clearly visible tidal arms (Ibata et al. 1997, Majewski et al. 2004a and in preparation).

(4) *Are there extended populations of dSph stars?* If dSphs are in tidal disruption, then we should find stars outside their nominally bound centers. Evidence for extended populations (i.e. beyond the nominal King limiting radii, r_{lim}) of most of the dSph galaxies was found in all eight dSphs surveyed photographically by Irwin & Hatzidimitriou (1995); but the issue of photometrically detected populations beyond r_{lim} of various of the dSphs has remained a controversial subject (e.g., see summaries in Majewski et al. 2004b, Westfall et al. 2004). This is difficult work because the surface brightnesses of these features are 31 mag arcsec^{-2} or less (recall that the orders of magnitude brighter *cores* of most of the dSphs were only discovered in the past half-century!), so the signals can easily be swamped by contaminating back/foreground sources. Recently, however, using 1 through 10-m telescope spectroscopy, we have proven the existence of velocity members outside the r_{lim} of Carina (Majewski et al. 2004b, Muñoz et al. 2004), Sculptor (Westfall et al. 2004), Leo I (Sohn et al. 2004), Ursa Minor (in preparation), and, of course, the more nearby Sgr dSph (Majewski et al. 2004a), with the aid of techniques for efficiently pre-selecting these stars (§2). In the case of Carina, we have found velocity members more than $3.3r_{lim}$ away from the center. To keep such stars bound requires $M/L > 650$ for Carina! However, we demonstrate (Muñoz et al. 2004) that it is more likely that these stars represent unbound, tidal debris. As Majewski et al. (in preparation) demonstrate, most Galactic dSphs have structural and dynamical similarities that in the Sgr system are the direct result of tidal disruption. This is certainly interesting evidence to suggest that tidal disruption is a natural state of the Galactic dSphs.

(5) *Do dSph galaxies build the Milky Way halo?* Clearly, with the example of the disrupting Sgr system, the answer is yes — and the process is ongoing. But how much has accretion of satellites at later times contributed compared to a primordial formation scenario? The answer would become more obvious were the halo to be thoroughly surveyed for substructure. Previous work by Majewski et al. (1996, 1999) suggests that in fact the halo may be entirely derived from accretion of smaller star systems.

(6) *What is the connection of dSphs to the predicted Cold Dark Matter (CDM) substructure?* CDM models that build galaxies hierarchically are already apparently overproducing satellites in the halos of Milky Way-like systems, and *not* interpreting the dSphs as CDM subhalos would only exacerbate the "missing satellites" problem of these models. Moreover, it is clear that at least some dSphs *are* being accreted by the Milky Way. Apart from providing subhalo candidates, these accreting satellites create dynamically cold tidal streams that are extremely useful probes of the lumpiness of the

halo potential (e.g., Johnston et al. 2002). When applied to this problem (Majewski et al. 2004a), the very long Sgr tidal stream is consistent, however, with a Milky Way halo that is actually rather smooth (though lumpier halos cannot yet be definitively ruled out).

TARGET SELECTION AT LOW SURFACE DENSITIES

A major problem confronting studies of dSph galaxies, particularly in outer regions where the effects of tidal disruption would be most obvious, is that the sky density of targets with brightnesses amenable to current spectroscopy — generally the red giants — is typically very low. For example, our work on Carina, Sculptor and Ursa Minor demonstrates that the number of $r > r_{lim}$ giant candidates within a few magnitudes of the giant branch tip quickly drops to a few per deg^2, which is a few percent of all stars to the corresponding apparent magnitude even at high Galactic latitudes. Even were one to have a multi-object spectroscopic capability, one would need to target hundreds of stars in a square degree to find the few stars of interest (Fig. 1) — not an efficient use of large telescope time. Various groups have been careful to pre-screen stellar sources for those that lie in the color-magnitude diagram (CMD) on/near the dSph's giant branch. But, as Figure 1 demonstrates, this generally improves the efficiency by only a factor of a few, so that the amount of contaminating sources still well outnumber the dSph stars (for example, in the Sculptor example shown, by a factor of ten at the King limiting radius).

For our work on the dSphs mentioned above, we have improved sample reliabilty an order of magnitude further by not only screening stars in the CMD, but by also pre-filtering for those stars most likely to be giants. This giant-filtering is based on imaging in the *DDO*51 intermediate band filter, which samples the stellar flux in the region of the gravity-sensitive Mgb+MgH features at 515 nm (see Majewski et al. 2000). Washington *M* filter images serve as "continuum". By suppressing the contamination of foreground dwarf stars to nearly an order of magnitude *below* the dSph giant density, this scheme has allowed us to winnow our candidate lists to samples found spectroscopically to be "extratidal" dSph stars at the >50%, and often approaching the 100%, reliability level. With this methodology we have probed the extended populations of these dSphs well beyond r_{lim}, and, in the case of Ursa Minor we have verified dSph members as much as $5r_{lim}$ from center. Thoroughly vetting target lists before heading to Keck, Gemini, Magellan, VLT, HET, SALT, etc. translates to optimal use of these valuable resources.

FUTURE WORK WITH HET/SALT-LIKE TELESCOPES

As work with the Sgr system has demonstrated, there are immense scientific rewards — an improved understanding the structure and evolution of dwarf satellite galaxies *and* the Milky Way — if one can track the "extratidal" populations of satellites over large angles of the sky. It would also be desirable to undertake a general survey of halo substructure beyond that which can be tracked from presently known satellites. But the choice of stellar tracers is critical. The outer halo — to 100 kpc or more — is of particular interest because this is the typical distance of the currently known dSph satellites and halo substructure at these radii should be less dynamically mixed. Metal-

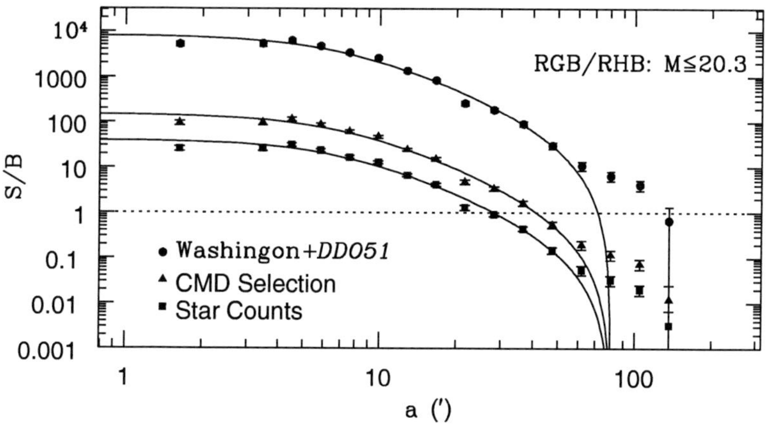

FIGURE 1. Comparison of relative dSph star signal, S, to contaminating fore/background stars, B, as a function of radial distance, a, from dSph center for various target selection strategies (in this case, data derived from the Sculptor study by Westfall et al. 2004). The bottom points/curve correspond to the actual dSph sample content one would get for a "blind" sampling of all stars to Washington $M = 20.3$. The middle points/curve shows that even applying a CMD-filtering scheme will yield only at best 10% actual dSph stars outside the King limiting radius. Our method for using Washington+*DDO*51 photometry to also identify stars likely to be giants improves the success rate by 50× over CMD-filtering only.

poor K giants at these distances have $V \sim 18$. In comparison, a horizontal branch star of the same brightness will be less than ~ 40 kpc away, while MSTO stars of this apparent magnitude, though much more numerous, will only be ~ 5 kpc away. One can probe to fainter apparent magnitudes for these types of stars, but the required exposure times become very long (exceeding, e.g., the typical single pointing integration time limits of a SALT-like telescope): Generally, one needs to expose 100 times longer for MSTO than RGB stars at the same V for the same S/N. A substantial, systematic survey of the outer halo as envisioned can only hope to be effective with SALT using red giant stars. However, as shown above, the density of such stars in tidal streams is generally too low to take real advantage of multi-object spectroscopy with small field-of-view systems.

Nevertheless, in many ways HET/SALT-like telescopes are ideal instruments for the study of halo substructure because: (1) tidal debris streams (like those from the Sgr and Monoceros systems) can stretch hundreds of degrees, (2) multiple streams are expected to network the sky, while (3) the spacing and brightness of their constituent red giant members are well-matched to the cadence of single-slit, SALT spectroscopy. However, because such stars are far outnumbered by foreground disk dwarf stars, a key necessity before undertaking such a survey is a reliable list of giant star targets. We have discussed (§2) one method for efficiently producing such a list, and are presently using it to create the Grid Giant Star Survey (Patterson et al. 2001) — a pilot program to do a partially-filled, all-sky survey to find candidate halo streams of giant stars.

We are grateful support from the NSF, NASA, the David and Lucile Packard Foundation, and The F. H. Levinson Fund of the Peninsula Community Foundation.

REFERENCES

1. Ibata, R. A., Wyse, R. F. G., Gilmore, G., Irwin, M. J., & Suntzeff, N. B. 1997, AJ, 113, 634
2. Irwin, M., & Hatzidimitriou, D. 1995, MNRAS, 277, 1354
3. Johnston, K. V., Spergel, D. N., & Haydn, C. 2002, ApJ, 570, 656
4. Kleyna, J., Wilkinson, M. I., Evans, N. W., Gilmore, G., & Frayn, C. 2002, MNRAS, 330, 792
5. Kleyna, J. T., Wilkinson, M. I., Evans, N. W., Gilmore, G. 2004, MNRAS, 354, 66
6. Łokas, E., Mamon, G. & Prada, F. 2004, MNRAS, submitted (astro-ph/0411694)
7. Majewski et al. 2004a, ApJ, 128, 245
8. Majewski et al. 2004b, AJ, submitted
9. Majewski, S. R., Munn, J. A., & Hawley, S. L. 1996, ApJL, 459, L73
10. Majewski, S. R., Ostheimer, J. C., Kunkel, W. E., Johnston, K. V., Patterson, R. J., & Palma, C. 1999, IAU Symp. 190: New Views of the Magellanic Clouds, 190, 508
11. Majewski, S. R., Ostheimer, J. C., Kunkel, W. E., & Patterson, R. J. 2000b, AJ, 120, 2550
12. Muñoz, R.R., Majewski, S.R., Kunkel, W.E., Johnston, K.V., Frinchaboy, P.B., Law, D.R. & Patterson, R.J. 2004, in preparation.
13. Patterson, R. J., Majewski, S. R., Slesnick, C. L., & et al. 2001, ASP Conf. Ser. 246: IAU Colloq. 183: Small Telescope Astronomy on Global Scales, 65
14. Sohn, S., Majewski, S. R., Siegel, M. H., Ostheimer, J. C., Patterson, R. J., & Muñoz, R. R. 2004, in preparation
15. Westfall, K.B., Majewski, S.R., Ostheimer, J.C., Frinchaboy, P.M., Kunkel, W.E., Patterson, R.J. & Link, R. 2004, AJ, submitted

CONCLUDING REMARKS

SAWLT in Toruń

Brian Warner

Department of Astronomy, University of Cape Town, Rondebosch 7700, South Africa

Abstract. This is an idiosyncratic summary of the Toruń Workshop.

It is possibly no coincidence that the title of this Workshop provides an acronym that is phonetically the same as that of the large telescope being built in South Africa. Although the Workshop was of interest to any institution possessing, or pregnant with, a very large telescope, there was the expected emphasis on the Southern African Large Telescope in which Poland has a share. If the enthusiasm for SALT irritates (especially for medical reasons), think of it as Some Anonymous Large Telescope.

A CONTINUING LIFE FOR SMALL TELESCOPES

Although the Workshop was largely about large telescopes, it was a welcome relief to find that medium sized (and even small sized) telescopes were given honorable mention. With the tendency at some major observatories over the past two decades to close smaller telescopes when a large one has been acquired, in some countries there has been a diminishing pool of useful telescopes. Smaller telescopes (STs) are important in their own right, as educational tools and for observing the brighter objects that are not going to be allowed to become targets for the LTs. But it is also true that the way some LTs are used is approaching the way that satellite telescopes are managed, which means that users need not necessarily know anything at all about the real sky, and this is tending to reduce the use of STs as training camps for neophyte astronomers. In the context of LTs, however, STs have at least the following important roles to play:

a) **As wide field search machines to find and classify the objects that will become targets for LTs.** Examples of such surveys currently underway are the Sloan Digital Sky Survey and the Hamburg/ESO Survey. These use specialized dedicated STs. But existing STs that might otherwise be looked on deprecatingly can be fitted with tackle that converts them to respectable support instruments for LTs. An example of giving an old telescope a new ambition in life is the provision of a relatively wide field CCD camera and high speed photometric facility at the prime focus of the McDonald 82 in telescope (Nather and Mukadam [1]).
The forthcoming LAMOST telescope, described at the Workshop by Hu, is, by LT standards, a medium size telescope, but with a big heart. The ultimate aim of providing spectra for all stars brighter than $V = 16$, and in selected areas to $V = 18$, will undoubtedly reveal hundreds of stars requiring further attention by LTs.

b) **As low pass filters.** The astronomical output of the survey instruments will in

CP752, *Stellar Astrophysics with the World's Largest Telescopes: First International Workshop*
edited by J. Mikołajewska and A. Olech

many cases require further investigation, at least of the brighter objects, before committing LT time to their detailed study. Standard 2 - 4 m class telescopes should be preserved for this work.

c) **For astrometry.** Accurate positions are required for LT observations - these must be provided by smaller telescopes. Such positions are needed not only for finding targets - they are essential for preparing multi-slit or multi-fibre masks, templates or layouts, without having to use the LT for the purpose.

d) **For photometric calibrations.** A telescope like SALT, which has a varying effective aperture during observation and a tessellated mirror, is difficult to use for accurate fluxed spectral distributions. Close cooperation with STs able to provide wide band absolute calibrations will often be necessary.

PROGRAM SELECTION FOR SALT

In his introductory review, speaking from several years of experience with the Hobby-Eberly Telescope, Larry Ramsay listed the following important categories of objects that are suited to the strengths and weaknesses of telescopes like HET and SALT.

a) The target classes should ideally be distributed uniformly over the sky. This is easier to arrange in the northern sky than in the southern: the LMC and SMC are decidedly not uniformly distributed. However, by accident of design, SALT provides maximum continuous observational coverage (not quite a Continuous Viewing Zone as in the HST, but the next best thing - up to 5 hours with an azimuth shift in the middle) for the Magellanic Clouds, which are at its southernmost accessible declination.

b) An areal density of ∼few per square arc minute for HET (or ∼several for SALT, which has a wider usable field of view). This is what will give optimum performance when in survey mode.

c) Timescales of variation (photometric or spectroscopic) > 1 day or < 1 hour. This is predicated by the relatively small average amount of time that SALT can spend on any given object (again, things are rosier near the southern declination limit, which is good news for observers of the southern hemisphere's brightest dwarf nova - VW Hyi). The Workshop contribution by Phil Charles, on time domain astrophysics, drew attention to the special niche that SALT can occupy in studying the most rapid (<few secs) variations at high S/N. The SALTICAM instrumentation, which makes this possible, will be unique among LTs. In general discussion it was remarked that getting the most out of this SALT strength, which will include rapid response to unpredictable events, will need the provision of excellent communication between the applicant for Director's Discretionary Time, the Director, and the Duty Astronomer. This will go against the tenets of bureaucracy - which normally require forms filled in triplicate and cleft-stick technology.

d) Spectroscopic observations. As originally planned and named, HET was the Spectroscopic Survey Telescope - it only changed its name when it came into some money. As its title implied (and this is true, but to a lesser extent, for SALT, which has the additional high speed photometric capability), spectroscopy is expected to be the dominant occupation of telescopes like HET and SALT. In the case of the latter, the range and versatility of PFIS and the high stability of the HRS should furnish opportunities for the highest quality science.

CLASSES OF OBSERVATION

In his opening talk Larry Ramsey also defined a number of categories into which the most SALT and HET observational programs may be divided. These are Survey Programs, Synoptic Programs, Targets of Opportunity, and In-depth studies. The reader of the Proceedings of this Workshop can pass the various contributions through these sieves in order to see where each fits (noting, though, that some are descriptions of instrumentation rather than observational programs). Here I will illustrate just a few of them.

a) **Survey Programs.** The searches for extra-solar planets, as described by Niedzielski and by Liu, are examples of what the SALT HRS will be able to provide. The motivation for the (Niedzielski) HET survey of K giants leads to an unusual product - the hope is to find noise rather than signal. But we assume that those stars found to have companions, and hence to be unsuitable for the astrometric stability required for the SIM grid of reference stars, will not simply be discarded - one person's unwanted signal is still a worthwhile signal.
Not mentioned at the Workshop was the ability of SALT to push study of stars in existing surveys to fainter limits. A prime example is the Edinburgh-Cape survey of southern UV-rich stars, which produces candidates to $V \approx 18.0$ but for which follow-up observations have been largely limited to $V < 16.5$ (Stobie et al [2]). There are other southern hemisphere surveys for which complete spectroscopic coverage is lacking - indeed, it would be of value to accumulate a homogeneous set of low dispersion spectra of all known southern quasars, cataclysmic variables, select categories of white dwarfs, etc, compiled from the various extant catalogues. Intercomparison of such large numbers of spectra often leads to better classification schemes, discovery of new general categories and recognition of anomalous individuals.
If the Large Synoptic Survey Telescope (http://www.lsst.org/lsst_home.shtml) gets on the ground, it will provide an abundance of objects to survey and study in detail for decades to come.

b) **Synoptic Programs.** For synoptic programs the SALT queue-scheduling philosophy is ideal. Many presentations at the Workshop fall into this category - for example, spectra of Trans-Neptunian objects (Kwiatkowski), light curves and spectra of long period variables (Whitelock), CVs in the Magellanic Clouds (Shara). Most use of SALT would be expected to fall into the survey and synoptic categories, as has been the case for HET (Ramsey). Many programs are both synoptic and survey - the search for more CV outbursts in the Magellanic Clouds, for example.

c) **In-Depth Studies.** Again Q-scheduling is a strength and many examples were given at the Workshop, e.g., spectra of planetary nebulae (Corradi, Conçalves), symbiotic stars (Zhu, Quiroga), studies of binaries in clusters (Kaluzny). Harries pointed out in discussion that so much of modern cosmology depends on the use of supernovae as standard candles, so they are deserving of great observational attention, especially looking for the signature of asymmetry provided by linear polarization, which would then need to be included in models.

d) **Targets of Opportunity.** All unpredictable events fall into this category, of which examples are given by Gamma Ray bursts (Stanek) and CV outbursts (Shara). As discussed above, SALT has potential for being the most rapid response LT in the southern hemisphere.

I hope that the Proceedings of this Workshop will be of value not only to those in the final planning stages of using the SALT, but also to others using or contemplating the construction of LTs around the world.

ACKNOWLEDGMENTS

My research is supported by the University of Cape Town. I am grateful to the Nicolaus Copernicus Astronomical Center in Warsaw for additional support and for hospitality.

REFERENCES

1. R. R. Nather and A. S. Mukadam, *ApJ*, **605**, 846 (2004).
2. R. S. Sobie et al, *MNRAS*, **287**, 848 (1997).

APPENDICES

LIST OF PARTICIPANTS

1. Annuk Kalju, Tartu Obs., Estonia, *annuk@aai.ee*
2. Balega Yuri, Special Astrophysical Obs., Russia, *balega@sao.ru*
3. Buckley David, SALT/SAAO, South Africa, *dibnob@saao.ac.za*
4. Chaboyer Brian, Dartmouth College, USA, *chaboyer@heather.dartmouth.edu*
5. Charles Phil, SAAO, South Africa/Univ. of Southampton, *pac@asto.soton.ac.uk*
6. Cikała Marcin, Centre for Astron., N. Copernicus Univ., Poland, *Marcin.Cikala @astri.uni.torun.pl*
7. Corradi Romano, Isaac Newton Group of Telescopes, Spain, *rcorradi@ing.iac.es*
8. Czart Krzysztof, Centre for Astron., N. Copernicus Univ., Poland, *Krzysztof.Czart @astri.uni.torun.pl*
9. Dreizler Stefan, Univ.-Sternwarte Goettingen, Germany, *dreizler@astro.physik.uni-goettingen.de*
10. Frąckowiak Sylwia, Centre for Astron., N. Copernicus Univ., Poland, *sylwiah@as-tri.uni.torun.pl*
11. Galan Cezary, Centre for Astron., N. Copernicus Univ., Poland, *cgalan@astri.uni. torun.pl*
12. Gęsicki Krzysztof, Centre for Astron., N. Copernicus Univ., Poland, *gesicki@astri. uni.torun.pl*
13. Gonçalves Denise R., Instituto de Astronomia, Geofisica e Ciencias Atmosfericas - USP, Sao Paulo, Brasil, *denise@astro.iag.usp.br*
14. Górny Sławomir, N. Copernicus Astron. Center, Poland, *skg@ncac.torun.pl*
15. Grebel Eva K., Univ. of Basel, Switzerland, *grebel@astro.unibas.ch*
16. Gromadzki Mariusz, N. Copernicus Astron. Center, Poland, *marg@camk.edu.pl*
17. Harries Tim, Univ. of Exeter, UK, *th@astro.ex.ac.uk*
18. Hu Jingyao, National Astron. Obs., China, *hjy@class1.bao.ac.cn*
19. Kałużny Janusz, N. Copernicus Astron. Center, *jka@camk.edu.pl*
20. Kraus Michaela, Astron. Inst., Utrecht Univ., The Netherlands, *M.Kraus@phys.uu.nl*
21. Kwiatkowski Tomasz, Poznań Astron. Obs., Poland, *tkastr@vesta.astro.amu.edu.pl*
22. Liu Yujuan, National Astron. Obs., Chinese Acad. of Sci., China, *lyj@yac.bao.ac.cn*
23. Majewski Steven, Univ. of Virginia, USA, *srm4n@virginia.edu*
24. Mazeh Tsevi, Wise Obs., Tel Aviv Univ., Israel, *mazeh@wise.tau.ac.il*
25. Mikołajewska Joanna, N. Copernicus Astron. Center, Poland, *mikolaj@camk.edu.pl*
26. Niedzielski Andrzej, Centre for Astron., N. Copernicus Univ., Poland, *aniedzi@astri.uni.torun.pl*
27. Olech Arkadiusz, N. Copernicus Astron. Center, Poland, *olech@camk.edu.pl*
28. Osiwała Jędrzej, Centre for Astron., N. Copernicus Univ., Poland, *sk@astri.uni. torun.pl*
29. Pietrukowicz Paweł, N. Copernicus Astron. Center, Poland, *pietruk@camk.edu.pl*
30. Pułecka Marta, N. Copernicus Astron. Center, Poland, *mara@ncac.torun.pl*

31. Quiroga Claudio Hugo Martin, Facultad de Ciencias Astronomicas y Geofisicas - Universidad Nacional de La Plata, Argentina, *claudioq@muchi.fcaglp.unlp.edu.ar*
32. Ramsey Larry, HET/Pennsylvania State Univ., USA, *lwr@astro.psu.edu*
33. Rutkowski Artur, N. Copernicus Astron. Center, Poland, *rudy@camk.edu.pl*
34. Sarna Marek J., N. Copernicus Astron. Center, Poland, *sarna@camk.edu.pl*
35. Shara Michael, American Museum of Natural History, USA, *mshara@amnh.org*
36. Siódmiak Natasza, N. Copernicus Astron. Center, Poland, *siodmiak@ncac.torun.pl*
37. Siwak Michal, Astron. Obs., Jagiellonian Univ., Poland, *siwak@nac.oa.uj.edu.pl*
38. Stachowski Grzegorz, N. Copernicus Astron. Center, Poland, *gss@camk.edu.pl*
39. Stanek Krzysztof, Harvard-Smithsonian CfA, USA, *kstanek@cfa.harvard.edu*
40. Szczerba Ryszard, N. Copernicus Astron. Center, Poland, *szczerba@ncac.torun.pl*
41. Szostek Anna, N. Copernicus Astron. Center, Poland, *asz@camk.edu.pl*
42. Szpanko Marcin, Dept. of Astrophysics and Theoretical Physics, Opole Univ., Poland, *szpanko@uni.opole.pl*
43. Tobin William, Dept of Physics and Astron., Univ. of Canterbury, New Zealand, *william.tobin@canterbury.ac.nz*
44. Warner Brian, Univ. of Cape Town, South Africa, *warner@physci.uct.ac.za*
45. Wegner Gary, Dartmouth College, USA, *gaw@bellz.dartmouth.edu*
46. Whitelock Patricia, SAAO, South Africa, *paw@saao.ac.za*
47. Xu Yan, National Astron. Obs., Chinese Acad. of Sci., China, *xuyan@bao.ac.cn*
48. Zhao Gang, National Astron. Obs., Chinese Acad. of Sci., China, *gzhao@bao.ac.cn*
49. Zhu Zhen-Xi, National Astron. Obs., Chinese Acad. of Sci., China, *zhu@bao.ac.cn*

Author Index